Strategic Studies Institute
and
U.S. Army War College Press

RETURN OF THE BALKANS: CHALLENGES TO EUROPEAN INTEGRATION AND U.S. DISENGAGEMENT

Janusz Bugajski

May 2013

*****

Comments pertaining to this report are invited and should be forwarded to: Director, Strategic Studies Institute and U.S. Army War College Press, U.S. Army War College, 47 Ashburn Drive, Carlisle, PA 17013-5010.

*****

This manuscript was funded by the U.S. Army War College External Research Associates Program. Information on this program is available on our website, *www.StrategicStudiesInstitute.army.mil*, at the Opportunities tab.

*****

Research for this monograph was conducted partly in Washington, DC, including a thorough appraisal of relevant literature on both sides of the Atlantic. It also entailed regular visits to the Western Balkan countries for meetings with local analysts and government officials. In particular, I would like to thank Besian Bocka, my former administrative and research assistant at the Center for Strategic and International Studies in Washington, DC; and Dovile Sukyte, my former research assistant, for their invaluable help by providing me with voluminous amounts of reading material.

*****

*****

The Strategic Studies Institute and USAWC Press publishes a monthly e-mail newsletter to update the national security community on the research of our analysts, recent and forthcoming publications, and upcoming conferences sponsored by the Institute. Each newsletter also provides a strategic commentary by one of our research analysts. If you are interested in receiving this newsletter, please subscribe on the SSI website at *www.StrategicStudiesInstitute.army.mil/newsletter/.*

ISBN 1-58487-570-4

CONTENTS

FOREWORD

This Letort Paper assesses the prospects for further turbulence and conflict in the Western Balkans and weighs the implications for U.S. policy and for potential future military engagement. Although the region has slipped off the American radar screen in recent years, several unresolved disputes have the potential of escalating. This Paper systematically describes numerous causes of domestic and regional tensions and outlines a number of conflict scenarios.

Regional disputes are evident over the status of specific territories, the validity of administrative borders, the credibility of specific governments, and, in some cases, over the legitimacy of statehood itself. Democratic progress is difficult where state building is incomplete and contested. Furthermore, as the author underscores, incomplete, conflicted, and contested states present serious challenges for European Union (EU, or the Union) enlargement and the institutional absorption of the Western Balkans.

The region can become a gray zone where limited progress in implementing reforms is followed by prolonged periods of stagnation or even reversal. Such conditions provide fertile terrain for political and nationalist extremism and heighten exposure to destabilizing foreign influences. Although these are unlikely to generate extensive armed conflicts, as witnessed in the 1990s, they will create pockets of insecurity and violence that would disqualify several states from the prospect of EU membership. Such exclusion would, in turn, prolong and exacerbate local disputes.

At the same time, the soft power capabilities of the EU are weakening for a number of reasons, including resistance among member states to further enlargement given the Union's economic problems; disappointment with the performance of recent members Romania and Bulgaria, as well as with older member states such as Greece, Italy, and Spain, which are encumbered by massive sovereign debts; and the unfulfilled commitments of several Western Balkan aspirants in their quest for EU accession.

Europe's overall economic downturn will also have negative consequences for the Western Balkans. It will curtail investment and credits in the region, encourage EU enlargement exhaustion, and reinforce Balkan reform fatigue. It will also create space for populists and nationalists, who will benefit from economic stagnation and public anger to promulgate ethno-nationalist solutions to mounting domestic challenges. Economic hardship decreases trust not only in incumbent governments but also in democratic institutions and international agencies.

Such negative scenarios would place the onus on key international actors to find credible solutions. However, the EU's effectiveness as an institution builder is coming under increasing scrutiny at a time when the United States is preoccupied with more pressing crises outside Europe. While Washington has spent the last decade extricating itself militarily and politically from the Western Balkans and allowing EU institutions to assume the leading role, unresolved disputes that are mishandled by an indecisive and divided Union could pull Washington back into the region. This could be evidenced in more intensive diplomacy and intrusive mediation or even in the context of new peacekeeping missions.

The Paper concludes by offering a number of concrete recommendations for the U.S. administration, European governments, international institutions, and local political leaders to avoid the dangerous pitfalls of state paralysis, territorial fracture, and regional destabilization. In particular, policies must be geared toward preventing a scenario whereby U.S. ground forces are called upon to participate in renewed peacemaking operations. The priorities must include more comprehensive strategic intelligence gathering, the identification and monitoring of local and foreign political actors promoting instability, early warning signals that can pinpoint and defuse impending conflicts, a strong Allied diplomatic response to any deterioration of political conditions, and a firmer transatlantic strategic commitment to bringing all countries in the region into the North Atlantic Treaty Organization and the EU.

DOUGLAS C. LOVELACE, JR.
Director
Strategic Studies Institute and
U.S. Army War College Press

ABOUT THE AUTHOR

JANUSZ BUGAJSKI is a policy analyst, writer, lecturer, and television host based in the United States. His current positions include Senior Associate at the Europe Program at the Center for Strategic and International Studies; Chair of South-Central Europe Area Studies at the Foreign Service Institute, U.S. Department of State as an independent contractor; and host of the "Bugajski Hour," a monthly program broadcast on Albanian Screen television from Tirana, Albania. Mr. Bugajski served as a consultant for the U.S. Department of Defense, the U.S. Agency for International Development, the International Research and Exchanges Board, and other U.S. organizations. He has lectured at the Smithsonian Institution, the Woodrow Wilson Center, the National Defense University, and at numerous American universities and has testified before various congressional committees. Mr. Bugajski is a regular contributor to various U.S. and European newspapers and international journals; he is a columnist for newspapers in Albania, Bosnia-Herzegovina, Bulgaria, Croatia, Georgia, Kosova, and Ukraine. He maintains professional contacts with research institutions throughout the region. Mr. Bugajski is the author of *Journeys of No Return: The Balkan Sagas of Anvil Kutlas* (Create Space, 2011); *Georgian Lessons: Conflicting Western and Russian Interests in the Wider Europe* (Center for Strategic and International Studies [CSIS], 2010); *Dismantling the West: Russia's Atlantic Agenda* (Potomac, 2009); *America's New European Allies* (Nova Science, 2009); *Expanding Eurasia: Russia's European Ambitions* (CSIS, 2008); *The Eastern Dimension of America's New European Allies* (U.S. Army War College, 2007); *Kosova: From Occupation to Independence* (Koha,

Prishtina, 2006); *America's Atlantic Bridges: America's New European Allies*, with Ilona Teleki, (Rowman & Littlefield, 2007); *Cold Peace: Russia's New Imperialism* (Praeger/Greenwood, 2004); *Political Parties of Eastern Europe: A Guide to Politics in the Post-Communist Era* (M. E. Sharpe, 2002); *Toward an Understanding of Russia: New European Perspectives* (ed.) (Council on Foreign Relations, 2002); *Ethnic Politics in Eastern Europe: A Guide to Nationality Policies, Organizations, and Parties* (M. E. Sharpe, 1994); *Nations in Turmoil: Conflict and Cooperation in Eastern Europe* (Westview, 1995); *East European Fault Lines: Dissent, Opposition, and Social Activism* (Westview, 1989); and *Czechoslovakia: Charter 77's Decade of Dissent* (Praeger/CSIS, 1987).

SUMMARY

Political developments in the Western Balkans with a direct impact on regional security must be closely monitored. In several countries, disputes continue over the validity of administrative borders and the credibility of specific governments, and, in some cases, over the legitimacy of statehood itself. Democratic progress becomes problematic where state building is incomplete and contested. Furthermore, internally conflicted and externally contested states present challenges for European Union (EU) integration, where the focus of the EU must not only be on democratic consolidation and economic reform but also on institutional legitimacy, state building, and regional security.

Economic stagnation throughout the EU will have negative consequences for the Western Balkans. It curtails investment and credits, raises opposition to further EU enlargement, and reinforces Western Balkan reform fatigue. Such developments provide space for populists and nationalists who will benefit from economic stagnation and public frustration and trumpet xenophobia as a solution to mounting domestic challenges.

The political status quo controlled by entrenched parties coupled with growing economic inequalities, a lack of sufficient judicial reform, the pervasiveness of official corruption, a sense of injustice, and unfulfilled economic and occupational expectations among citizens all have a negative impact on stability. They deepen public alienation, demoralization, resentment, and anger; increase crime and lawlessness; provide ammunition to new protest movements; and encourage extremism. Such sentiments can be turned against ethnic and religious minorities or toward broader causes such as border revisions and territorial acquisitions that fuel conflicts with nearby states and with international players.

The Western Balkan region has acquired a proliferation of precedents regarding national and territorial self-determination, autonomy, and secession that could be adopted by current or aspiring ethno-national leaders. Minority representatives in a number of states have viewed the fracturing of Yugoslavia and the independence of seven of its eight federal units as potentially repeatable precedents. Moreover, renewed conflicts over territory will be generated if domestic ethnic turmoil becomes increasingly interconnected among neighboring countries.

The EU's effectiveness as a promoter of reform will come under growing scrutiny if economic prospects in the Western Balkans diminish and disillusionment with the Union as an institutional destination increases. Although EU membership is not the panacea for resolving every conflict and eliminating all negative trends in the Western Balkans, the credible and timely prospect of accession into the Union helps keep democratic reforms on track as conditions for entry. Without such reforms, much of the progress achieved in the Western Balkans since the end of the Wars of Yugoslav Succession (1991-99) can stall or even unravel, and U.S. disengagement from the region will appear premature.

The EU has entered a period of prolonged economic uncertainty, social turmoil, political dispute, and institutional confusion. While the Union's limitations as a hard power have been evident in its disjointed foreign policies and restricted military capabilities, its political and economic model may also fade as an instrument of attraction if it closes its doors to further enlargement and if the Union begins to splinter in the midst of the expanding economic crisis.

In recent years, the United States has been preoccupied with pressing priorities outside Europe and now plays a diminished role in the Western Balkans. While Washington has spent the last decade extracting itself from the region and allowing EU institutions to assume the leading role, renewed conflicts that are mishandled by an indecisive and divided EU could again pull the United States back into the region in dealing more directly with a spiral of instability.

A number of policies must be pursued to prevent a scenario whereby America's European partners call upon U.S. ground forces to participate in peacemaking or peacekeeping missions in the Western Balkans. Such priorities must include more comprehensive strategic intelligence gathering, the identification of local and foreign political actors promoting instability, early warning signals regarding impending conflicts, a strong diplomatic response to any deterioration of political conditions in each Balkan state, and a firmer U.S. and Allied strategic commitment to bringing all countries in the region into the North Atlantic Treaty Organization and the EU.

October 2012

RETURN OF THE BALKANS: CHALLENGES TO EUROPEAN INTEGRATION AND U.S. DISENGAGEMENT

I. INTRODUCTION: UNCERTAIN FUTURES

For the first time in its modern history, the entire Balkan Peninsula has the opportunity to coexist under one security and developmental umbrella combining the North Atlantic Treaty Organization (NATO) and the European Union (EU). Unfortunately, European and American leaders have been unable to complete such a unique historic vision, while the progress of several Western Balkan countries continues to be undermined by a plethora of political, social, economic, and ethno-national tensions and disputes. This monograph focuses on the escalating security challenges facing the Western Balkans, assesses the shortcomings and deficiencies of current international engagement, considers future prospects for U.S. military involvement, and offers recommendations for curtailing conflict and promoting the region's international institutional integration.

The Failed State Index, an annual ranking of 177 states based on their levels of domestic stability and institutional capacity, places Bosnia-Herzegovina, Serbia, Kosova, and Macedonia within the "warning" category.[1] In these countries, social, economic, and political indicators highlight the potential for turmoil and disruptions in the functioning of the state with renewed bouts of inter-communal tensions, violence, and possible insurgency.

Although EU membership is not the panacea for resolving all conflicts and negative trends in the Western Balkans, the credible and timely prospect of accession into the Union helps keep democratic reforms on track as conditions for EU entry. Without such reforms, much of the progress achieved in the Western Balkans since the end of the Wars of Yugoslav Succession (1991-99) can stall or even unravel, and U.S. disengagement from the region will turn out to be premature. Moreover, renewed conflicts will be generated if domestic turmoil with ethnic dimensions in several Balkan states becomes increasingly interconnected.

The EU's limitations as a hard power have been evident for many years in its disjointed foreign policies and restricted military capabilities. However, its political and economic model may also be fading as an instrument of attraction if it closes its doors to further enlargement or indefinitely delays prospects for new members. A sentiment of skepticism within the EU toward further enlargement has grown among EU publics during the era of austerity and as the budgetary and debt crisis have propelled several Mediterranean countries toward prolonged economic uncertainty that generates social and political turmoil. Even more ominously, if the Union itself begins to splinter in the midst of its protracted economic crisis, the possibility of institutional closure may leave the Western Balkan states stranded.

The EU's soft power capabilities are waning for several reasons, including internal resistance among member states to further enlargement, given the Union's relentless economic and financial problems; disappointment with the performance of recent members Romania and Bulgaria, as well as with older members such as Greece, Italy, and Spain, which are encumbered by massive sovereign debts; and the unfulfilled commitments of several Western Balkan aspirants in their quest for EU accession. Europe's economic downturn will have negative consequences for the Western Balkans. It will curtail investment and credits in the region, encourage EU enlargement exhaustion, and reinforce Western Balkan reform fatigue. It will also create space for populists and nationalists, who will benefit from economic stagnation and public anger and promulgate ethno-nationalism and xenophobia as solutions to mounting domestic challenges.

As economic prospects in the Western Balkans are likely to diminish in the near term, disillusionment with the EU as an institutional destination may also spread. The Union's effectiveness as an institution builder and a factor of stability will come under greater scrutiny. This can occur at a time when the United States is preoccupied with more pressing crises outside Europe and calculates that it can only play a secondary role in the Balkans. Increasingly interconnected conflicts will undermine reformist leaders and once again raise group identity and ethno-nationalism to the forefront. Recession and economic hardships decrease trust not only in incumbent governments but also in democratic institutions and international agencies.

Factors with a direct impact on Western Balkan security need to be closely monitored. Regional disputes remain active over the status of specific territories, the validity of administrative borders, the credibility of specific governments, and, in some cases, over the legitimacy of statehood itself.[2] Democratic progress is difficult where state building is incomplete and contested. Furthermore, incomplete, conflicted, and contested states present challenges for EU integration, where the focus of the Union must not only be on democratic consolidation and economic reform but also on state building, institutional legitimacy, and regional stability.

For example, Bosnia-Herzegovina remains a divided state whose legitimacy continues to be questioned internally. Meanwhile, the EU is hesitant to press for solutions without consensus among leaders of the three major ethno-national groups. This position has perpetuated the political and institutional stalemate that blocks EU accession. At the same time, the EU has failed to upgrade Kosova to the status of a contractual partner with prospects for Union entry.[3] Five EU members do not recognize Kosova's statehood, and EU representatives are fearful of alienating Serbia and stimulating greater regional instability.

The Western Balkans have acquired a proliferation of precedents regarding national self-determination, territorial autonomy, and outright secession that could be adopted by current or aspiring ethno-national leaders. Minority representatives in a number of states have viewed the fracturing of Yugoslavia and the independence of seven of its eight federal units as potentially repeatable precedents. Moreover, renewed conflicts will be generated if domestic turmoil with ethnic dimensions becomes increasingly interconnected among neighboring countries.

For instance, the independence of Kosova, a formerly autonomous region of Serbia that also possessed federal status in Yugoslavia, has been claimed as a usable precedent by the autonomous Serb entity in Bosnia-Herzegovina and potentially by other minority enclaves in the region. The opinion of the International Court of Justice (ICJ) that Kosova's declaration of independence did not conflict with international law has given ammunition to other separatist leaders and regions, even though the ICJ did not support Kosova's right to independence.[4] However, in such cases, perceived precedents are more important than the implications of international law.

A plausible partition of Kosova, with four northern municipalities containing a Serbian majority that does not recognize the legitimacy of Kosova's independence and decides to detach from the state, could spark demands for similar separation in Macedonia, Bosnia-Herzegovina, Serbia, and Montenegro. Even if a territorial division was conducted peacefully and by mutual agreement between the governments in Belgrade and Prishtina, it could encourage other minority populations to follow the Kosova Serb example. Political intransigence and the creation of parallel and separate administrative structures would be designed to convince international actors that a joint state is not feasible and ultimately destabilizing. Partition would be presented as the only viable option even though such territorial dismemberment would not comply with any administrative borders inherited from the former federal Yugoslavia.

In sum, the region can descend into a gray zone where splutters of progress in pursuit of reform are followed by prolonged periods of stagnation or even reversal. Such conditions provide fertile terrain for varieties of political radicalism and nationalist extremism and heighten exposure to destabilizing foreign influences. Although these are unlikely to generate extensive armed conflicts, as witnessed in the 1990s, they will create pockets of insecurity and violence that can disqualify several states from any immediate prospect of EU membership. Such exclusion would, in turn, prolong and exacerbate local disputes.

These negative scenarios would place the onus on key international actors. While the United States has spent the last decade gradually extricating itself militarily and politically from the region and allowing EU institutions to assume the leading role, unresolved disputes that are mishandled by an indecisive and divided Union could pull Washington back into the region by having to deal more directly with a spiral of instability. This could be evident in more intensive diplomacy and intrusive mediation or even in the context of new peacekeeping missions.

Several urgent policies need to be pursued by Washington to prevent a scenario whereby America's European partners call upon U.S. ground forces to participate in renewed peacemaking or peacekeeping operations in the Western Balkans. These priorities must include more comprehensive strategic intelligence gathering, the identification and monitoring of local and foreign political actors promoting instability, early warning signals that can pinpoint and defuse impending conflicts, a strong Allied diplomatic response to any deterioration of political conditions or inter-ethnic relations in each Balkan state and at the interstate level, and a firmer transatlantic strategic commitment to bringing all countries in the region into NATO and the EU.

This monograph assesses the prospects for further turbulence and conflict in the Western Balkans and the implications for U.S. policy and potential future military engagement. It focuses in turn on Serbian aspirations, the Bosniak national revival, regional Albanianism, the Macedonian impasse, the shortcomings of the EU, and the impact of U.S. and NATO downsizing in the region. It subsequently summarizes the numerous causes of domestic and regional disputes and outlines several negative consequences and conflict scenarios. The analysis concludes by offering a number of prescriptions and recommendations for the U.S. administration, European governments, international institutions, and local political leaders to avoid the dangerous pitfalls of state paralysis, territorial fracture, and regional destabilization.

II. SERBIAN ASPIRATIONS

In December 2011, Serbian President Boris Tadic issued a warning that the country could again sink into nationalism if there were long delays in Serbia's progress toward EU membership.[5] Although the message was received with some skepticism as a ploy to convince Brussels to give Serbia EU candidate status, it also indicated some anxiety about the country's political undercurrents.

In its progress report to the European Council on October 12, 2011, the European Commission recommended that Serbia be granted the status of an EU candidate country. However, the report also concluded that Belgrade's accession process should be withheld until it complied with EU requirements over Serbia's former province of Kosova by achieving "significant progress" in normalizing relations with the government in Prishtina. This would entail Belgrade's cooperation with the EU's rule of law (EULEX) mission in Kosova, respect for Kosova's territorial integrity, and agreement regarding Prishtina's participation in all regional forums.

The Serbian authorities did not meet all the EU stipulations. Nevertheless, with the Tadic government staking its political future on Union accession and arguing that any pre-election concessions over

Kosova would provoke a nationalist backlash, EU capitals became anxious that Serbia's exclusion could give fresh ammunition to radical parties. As a result, Brussels, Belgium, granted Serbia EU candidate status on March 1, 2012, on the eve of its general elections. However, this decision proved insufficient to diminish public support for a more nationalist option in both the presidential and parliamentary ballots.

Nationalist Undercurrents.

The presidential victory of former ultra-nationalist Tomislav Nikolic on May 20, 2012, injected a dose of uncertainty about Serbia's political development. In addition, on July 27, 2012, Ivica Dacic, the Socialist Party leader and former spokesman for former dictator Slobodan Milosevic, was sworn in as Prime Minister in the new government, stoking fears that Belgrade could adopt a more assertive nationalism in its foreign policy.

Several regional leaders refused to attend the inauguration of President Nikolic due to his controversial past and inflammatory comments after the elections.[6] Slovenian President Danilo Tuerk, the Chairman of Bosnia-Herzegovina's Presidency Bakir Izetbegovic, and Croatian President Ivo Josipovic boycotted the ceremony. Nikolic sparked controversy with his post-election statements denying that genocide had been perpetrated in Srebrenica in July 1995 and calling the Croatian town of Vukovar a Serbian city to which displaced Croats should not return.

Three major dangers face Serbia at its most testing time since the fall of Milosevic in October 2000: economic decline, political radicalization, and international isolation. Nikolic competed for the presidency primarily over prescriptions to rescue a faltering economy. The majority of impoverished citizens voted against the incumbent President Boris Tadic, believing that Nikolic could alleviate their material distress. His Progressive Party vowed to invest in agriculture and industry and increase taxes on the rich to fund state pensions. However, such propositions may prove to be unrealistic in practice.

Serbia's economic performance is deteriorating and a mood of public frustration and disenchantment is spreading. Unemployment stands at around 25 percent and is rising, having almost doubled in 3 years, while gross domestic product (GDP) growth is projected at a mere 0.5 percent in 2012 after falling below 2 percent in 2011. Meanwhile, Serbia's foreign debt has reached 24 billion euros and is steadily climbing. Some local observers believe that Serbia could follow Greece toward insolvency and a potential social explosion as high unemployment invariably drives populism and youth radicalism.

Despite his bold election pledges, President Nikolic has little room for maneuver during the Europe-wide economic downturn. The new leadership will be expected to negotiate with the International Monetary Fund (IMF) to unfreeze much-needed funding. Meanwhile, the EU will review Serbia's qualifications to begin projected accession talks. However, both EU and IMF conditions for Belgrade will necessitate financial discipline, including public-sector wage freezes, redundancies, and even pension cuts. Moreover, any major tax increases for entrepreneurs, as proposed by the new President, are likely to scare off foreign investors. In fact, Serbia's economic conditions are likely to deteriorate further throughout 2013.

Politically, whereas a broad coalition government seemed more likely to keep Serbia's reforms on track, it would enable Belgrade to better prepare for accession talks with the EU, resume the suspended dialogue with Prishtina, and implement the agreements already reached with Kosova, the more narrow Progressive-Socialist coalition, led by Prime Minister Ivica Dacic, could prove less accommodating.[7] This may exacerbate political polarization between reformist and nationalist camps and further divide society at a time of growing economic frustration.

Economic distress, political division, and social turmoil can inject stronger doses of nationalism into foreign policy. During the election campaign, Nikolic promised that Serbia would not stray from its "European path." But remaining on this path will prove problematic if the new President exploits the status of both Kosova and Bosnia-Herzegovina to score domestic political points. Moreover, such an approach would encourage Serbian leaders in northern Kosova and in Bosnia's Republika Srpska (RS) to harden their positions and press more resolutely toward autonomy and even secession.

Serbia's assertiveness would, in turn, provoke nationalist responses among neighboring states suffering from their own economic problems and political disputes. It can also isolate Serbia, and instead of remaining on the European path, Nikolic could reach toward Russia to buttress his presidency. Instructively, on his first foreign trip after the elections, President Nikolic was warmly received in the Kremlin where President Vladimir Putin asserted that the Serbs were Russia's "spiritual brothers."

In reality, relations between Moscow and Belgrade have been marked by mutual exploitation rather than solidarity. When Slobodan Milosevic captured the Serbian state and destroyed the Yugoslav federation in the early 1990s, he manipulated Russia to defend himself against Western pressure. He needed President Boris Yeltsin to demonstrate that Serbia was not alone, while carving up territories in neighboring republics to create an ethnically homogenous state. Yeltsin also needed Milosevic to prove that Russia remained a major power even though the Soviet Union had disintegrated and Moscow had lost its East European satellites. Belgrade played on Russia's superpower ambitions while Moscow exploited Serbia's mini-imperialist dreams.

However, unlike Milosevic who was manipulative toward Russia, Serbian nationalists appear to be more gullible. For instance, Nikolic once asserted that he would prefer to see Serbia as a Russian province rather than an EU member. When Nikolic was Speaker of the Serbian parliament, he also claimed that Moscow was bringing together nations to stand up against the hegemony of America and the EU.[8]

Former President Vojislav Kostunica's Democratic Party of Serbia is a pivot of the Eurosceptic bloc and will support the new government if it adopts a harder line on the international arena. Additionally, the hierarchy of the Serbian Orthodox Church (SOC) openly sides with nationalist and social conservative political groups.[9] For example, the head of the church, Patriarch Irinej, has backed the unification of Serb-inhabited lands in Bosnia-Herzegovina, Montenegro, and Kosova with Serbia. In seeking to reclaim its position as the "state religion," the church has been active in the media, army, and educational system, and is publicly outspoken about government policy.

However, following Irinej's election as patriarch in early 2010, differences have surfaced in the SOC hierarchy between a more modern stream that reaches out to other denominations and a traditional conservative institution represented by Bishop Artemije, the former head of the Raška-Prizren Diocese in Sandzak and Kosova. Artemije maintains close ties with the Russian Orthodox Church and appears supportive of the Kremlin agenda in the Balkans. Disputes between the two religious tendencies can also contribute to polarizing and radicalizing political discourse in Serbia.

Small ultra-nationalist groups have also emerged in Serbia in recent years, including Obraz with a clerical-monarchist agenda; the ultra-nationalist Serbian National Movement (SNP) 1389, led by Igor Marinkovic; the Ravna Gora movement that seeks to rehabilitate the World War II Cetnik leader Draza Mihailovic and upholds his Greater Serbia ideology; and the pro-Church organization Dveri Srbske. These groups view much of the Serbian political establishment as national traitors over such questions as Kosova and Bosnia's RS. They popularize their extremist ideas and recruit members primarily through the Internet and the social media.

Serbia vs. Kosova.

One of the most persistent obstacles to security and interstate cooperation in the Western Balkans is the stalled relationship between Serbia and Kosova. The two governments have diametrically differing positions on Kosova's statehood. While Prishtina has tried to leverage Washington and Brussels to gain more extensive international recognitions and enter the major multinational institutions, Belgrade's primary foreign policy objective has been to freeze Kosova's status and prevent its international inclusion.

Despite prolonged U.S. and EU involvement, little progress has been visible in forging a rudimentary partnership between the two countries. Moreover, the focus on gaining or precluding international recognition and integration has contributed to obstructing necessary economic and institutional reforms, constricted the process of state building in Kosova, and oriented international attention toward crisis management rather than regional development and Euro-Atlantic assimilation.

The launching of a dialogue between the governments of Serbia and Kosova, mandated by the United Nations (UN) General Assembly in September 2010, sponsored by the EU, and supported by the United States, was supposed to break the deadlock by unfreezing relations between the two capitals and developing points of cooperation. Although the talks were initiated in March 2011, their purpose and outcome has been subject to contradictory interpretations. They are unlikely to either eliminate Serbia's obstructive opposition to Kosova's sovereignty or provide impetus for transforming Kosova into a fully functioning state and member of international organizations such as the UN.

EU officials perceive the talks as a demonstration that disputes and practical problems between Serbia and Kosova can be resolved with the assistance of Brussels. U.S. officials generally share the EU perspective but would like to see the dialogue as a stepping-stone toward Kosova's recognition as an independent state by all European capitals, as this would help propel the country into international institutions.

Serbia's leaders have viewed the talks instrumentally both as a means for gaining EU candidate status and as a method for delaying any further recognitions for Kosova. The government in Prishtina has been much less sanguine about the discussions with Belgrade and was pressured into accepting the dialogue without the prospect of any tangible rewards. The Tadic administration appointed political director of the Foreign Ministry Borko Stefanovic as head of the delegation for the talks. Belgrade's team did not include senior government officials to minimize the appearance of legitimizing dialogue with a state that Serbia does not recognize.

President Tadic was prepared to meet with Kosovar Prime Minister Hashim Thaci despite allegations that the latter was involved in war crimes during the NATO-Serbia conflict. In December 2010, Dick Marty, a member of the Council of Europe (CoE) Parliamentary Assembly, issued a report alleging the killing of prisoners and the removal and illicit trafficking of human organs by members of the Kosova Liberation Army (KLA), in which Thaci held a senior position. The allegations remain under investigation, while Tadic stated that unless Thaci was formally indicted, he was prepared to meet with any credible Albanian representative without recognizing the legitimacy of the government in Prishtina. It will be instructive to see if President Nikolic adopts the same approach as his predecessor.

A second objective for Belgrade to engage in talks is to halt any further recognitions for Kosova by making these seem contingent on the long-term outcome of the dialogue. Belgrade does not consider Kosova's final status to be resolved and will studiously avoid acknowledging its statehood. A third government goal is to ease domestic pressure from nationalists who charge that officials have neglected the Kosova question. Although Kosova does not figure as a hot political issue on a daily basis, it can rebound against the authorities if Belgrade is perceived as surrendering its claims to the territory or making major concessions to Prishtina.

Foreign Minister Vuk Jeremic represented the less compromising stream within the Serbian administration and protected the government's flank from nationalist attacks. He will continue to campaign against Kosova's international recognition after being appointed President of the UN General Assembly in June 2012. Jeremic periodically launched initiatives to bring Kosova's statehood into question, including an International Court of Justice process to obtain a ruling on Kosova's declaration of independence, mobilizing the Yugoslav-era Non-Aligned Movement against separatism, appealing to Russia to back Serbia's diplomatic maneuvers, and focusing on the CoE investigations of Prime Minister Thaci to delegitimize the government in Prishtina.

The Tadic administration remained adamant that direct negotiations with Prishtina did not signal recognition of Kosova as an equal partner or a fully-fledged state. Instead, Belgrade's agenda for the talks involved a number of practical issues supported by the EU, including accounting for people missing following the 1999 war, resolving transport and telecommunications problems, and legalizing documents for Serbs living in Kosova.

Prishtina's agenda in the talks included the exchange of civil documentation such as the civil register and property records held by Belgrade; removing obstacles to cooperation in the Central European Free

Trade Area (CEFTA) and other regional initiatives where Belgrade blocked Prishtina's participation; ensuring freedom of movement through the recognition of passports, customs stamps, license plates, and driving licenses; the acceptance of school diplomas by both states; developing telecommunications and energy links; and the return of looted property.

Local analysts believe that Thaci pushed ahead with the dialogue to undo some of the damage to his credibility stemming from allegations of war crimes. The Marty report and the CoE investigations have been strongly criticized by all political parties in Kosova. There is a widespread supposition that the main purpose of the report was to delegitimize the new state and allow Serbian authorities to depict Albanians as co-responsible for war crimes. Opposition parties also urged the government not to start talks with Belgrade without the approval of the Kosova Assembly. Assembly speaker Jakup Krasniqi argued that parliament should adopt a resolution to determine the topics of the dialogue. His request was disregarded by the government as pressures mounted from the EU to launch the talks.

Washington stated that it would play a supportive role in the dialogue, with the EU setting the agenda. U.S. Ambassador to Kosova Christopher Dell declared that the talks should not tackle the question of Kosova's status or territorial integrity. The U.S. Department of State may have calculated that a more prominent American role could push Serbia into requesting that Russia join the dialogue, thereby creating potential rifts between Washington and Moscow.

Prishtina has been adamant on three questions: that the talks cannot call into question Kosova's final status; that Kosova's borders are inviolable; and that

Serbia must have no say in Kosova's administrative structure. In the optimum scenario, Prishtina calculates that the discussions may begin a process that will normalize relations with Serbia as an impetus for EU integration. Kosova's goal is for the dialogue to conclude with reciprocal recognition, the mutual acceptance of territorial integrity, and the affiliation of both countries inside the EU. In reality, EU officials have made no linkage between the talks and Prishtina's integration in the Union.

Kosova's political opposition challenged the legitimacy of any dialogue with Serbia without Belgrade's explicit recognition of Kosova's statehood. This criticism was buttressed by the constitutional crisis in April 2011 when Kosova's President Behgjet Pacolli had to resign because of procedural irregularities, and fresh elections were held in parliament for the head of state. The new administration planned to change the constitution in favor of direct presidential elections.

According to the Kosovar opposition, Belgrade and Prishtina should only conduct a dialogue as two distinct and mutually recognizing states; otherwise, the EU-sponsored talks favored Belgrade. Opposition leaders contend that the EU has become "status neutral" vis-à-vis Kosova in an effort to pull Serbia closer to the Union. Ultimately, the talks could delegitimize the Kosovar administration domestically, especially if Prishtina is pressed to make concessions that are seen to subvert statehood.

An agreement reached on February 24, 2012, between Belgrade and Prishtina regarding Kosova's representation in regional organizations unleashed a storm of controversy.[10] According to the deal, Kosova would be represented simply under the name "Kosovo," not the Republic of Kosova, and with a footnote that appeared to challenge the country's independence. The qualification read: "This label (Kosovo) does not prejudge the status of Kosovo and is in accordance with Resolution 1244 and the opinion of the International Court of Justice on Kosovo's declaration of independence." If the Kosova naming decision is perceived as a victory for Serbia in undermining Kosova's independence, it may serve to embolden rather than pacify nationalist demands on both sides.

Partition Proposals.

Kosova's partition is not the official policy of the Serbian government. Nonetheless, there are strong indications that this is Belgrade's ultimate objective. Some Serbian analysts believe that the bilateral talks may become a smokescreen for relinquishing Kosova, while gaining its northern districts for Serbia. This would not mean formal recognition of a rump Kosova as an independent state but rather an impetus for the eventual merger of a truncated Kosova with Albania and the absorption of northern Kosova by Serbia.

Serbia's promotion of Kosova's partition is based on the assumption that the EU is weak without intensive American involvement as well as by potential Russian support for Serbian secession.[11] In such a scenario, Belgrade would not surrender Albanian majority districts in southern Serbia as part of a potential territorial exchange with Prishtina. Instead, Serbian officials may be hoping for a future grand deal with Tirana over the heads of Kosovar leaders. In a statement to the media on March 18, 2011, President Tadic spoke about a "historic solution" between Serbs and Albanians and hinted that this could be an agreement to partition Kosova.[12] Current Prime Minister Ivica

Dacic openly stated on May 18, 2011, when he was still the Interior Minister, that the only possible compromise was Kosova's division.[13]

Approximately 60,000 Serbs live in northern Kosova alongside 4,000 Albanians and 3,000 members of other ethnic groups. Serbs form majorities in four municipalities—Leposavic, Zvecan, Zubin Potok, and Kosovska Mitrovica. Serbs and Albanians have little contact, as the latter live mostly in isolated villages and the two communities are separated in Mitrovica, the largest town.

While officially part of Kosova, the region has been under the de facto control of separate institutions funded by Belgrade, including town councils, health authorities, post offices, and schools. Several local Serbian organizations are openly hostile to the international presence and are considered illegal by Prishtina. They include the Union of Serbian Districts and District Units of Kosovo and Metohija, the Serbian National Council for Kosovo and Metohija, and a Serbian Assembly. Officials in Belgrade calculate that promoting parallel structures in the north will create a *fait accompli* for the creation of a Serbian autonomous region or for outright partition.[14]

During a visit to Belgrade on August 23, 2011, German Chancellor Angela Merkel demanded that the Serbian authorities assist in dismantling all parallel structures in northern Kosova. Belgrade claims it has little actual influence over the area and the minority has established its own institutions. The abstention of Serbs in the northern municipalities during Kosova's general elections and their boycott of Kosova's institutions pose the greatest challenge for Prishtina. This subregion has become the main source of instability and is almost entirely outside of Kosova's legal system.

The Kosova government accepted the internationally mandated Martti Ahtisaari plan, which stipulates elements of self-rule for the Serbian minority in cultural, educational, social, and other affairs. However, this has not satisfied Serbian activists who campaign for full separation and union with Serbia.

Prishtina recognizes extensive decentralization, but it will not countenance regional autonomy that could turn Kosova into another Bosnia-Herzegovina with two entities that paralyze the central government. Prishtina also rejects any form of northern autonomy with power-sharing arrangements between Belgrade and Prishtina in which Serbia would play a political role inside Kosova. Belgrade has also proposed a tripartite agreement with Prishtina and Brussels in which Serbia would "delegate" sovereignty to Kosova. The other two parties rejected the scheme, as it undermines Kosova's independence.

According to U.S. Assistant Secretary of State Philip Gordon, both "partition and land swaps are unacceptable solutions. If any such process is set in motion, there is no way that it can be confined to a single boundary line or that it can end peacefully."[15] Despite such warnings, the division of Kosova and exchange of territories have been discussed in both Prishtina and Belgrade. The question will continue to surface in the coming years. Kosovar Albanians are not prepared for any border concessions but may be more open to territorial exchanges with southern Serbian municipalities in the Presevo valley containing Albanian majorities.

Some analysts and officials privately assert that without stronger U.S. involvement in the Serbia-Kosova dialogue and in canvassing for Kosova's international recognition and membership of multinational institutions, the EU could veer from "status neutral" to "status negative" regarding Kosova's statehood. With Brussels in the lead role, partition could then be viewed as a viable option to placate Belgrade but may unwittingly intensify Albanian nationalism throughout the region.

Serbia's EU candidacy without commensurate progress by Kosova could create new obstacles for Prishtina by emboldening Belgrade to be more obstructive in future negotiations. Some EU officials and former representatives have voiced support for broad autonomy for northern Kosova, including Sweden's Foreign Minister Carl Bild.[16] Tadic backed such suggestions by stating that potential models for northern Kosova could be Northern Ireland within the United Kingdom or South Tyrol within Italy.[17] The former coordinator of the South East European Stability Pact, Erhard Busek, reportedly stated that an exchange of territory between Serbia and Kosova (Presevo valley for northern Kosova) might be the best solution for the region.[18] Other analysts have dismissed such proposals as stimulating further demands for autonomy and secession in several post-Yugoslav states.

Kosova's Violent Prospects.

Without credible international deterrents, the prospects for instability will accelerate in Kosova in the midst of economic distress and social turmoil. Numerous incidents of violence have occurred in recent years, including a standoff at two border crossings between Kosova and Serbia during the summer of 2011 when Prishtina wanted to establish its control by deploying units of the Kosova Police Force (KPF) but was thwarted by local squads of Serbian civilians. NATO eventually took over security along the border to prevent any escalation of violence while EULEX assumed control over the Jarinje and Brnjak border posts. The border crossings continued to be flash points for violence.[19] In early April 2012, Kosovo Forces (KFOR) peacekeepers informed Serb leaders that they planned to close illegal roads used to bypass the official border posts. Local Serbs blocked roads in the area for several months protesting over the presence of Kosova police and customs officials.[20]

On November 29, 2011, UN Kosova mission chief Farid Zarif warned that serious violence could erupt in the north at any time. This followed clashes between protesting Serbs and NATO forces that left 30 NATO soldiers and some 100 Serbs injured.[21] KFOR commanders issued alerts that conflicts could flare up either as a result of actions by the Kosova authorities or by groups of local Serbs.[22] This could happen either if the government in Prishtina attempted to establish full authority in the north or if Serbs initiated violence to provoke a crackdown and directly involve Belgrade in the ensuing crisis. Kosova's Interior Minister Bajram Rexhepi warned that force may be required to restore freedom of movement and the removal of barricades.

Local referenda or elections in northern Kosova outside of Prishtina's authority can also become a tripwire for violence. The four municipalities with a majority Serbian population held separate elections on May 30, 2010, unopposed by the international presence and with Prishtina powerless to intervene. Nationalist Serbian parties, including the Serbian Progressive Party and the Democratic Party of Serbia, dominate local politics in the region. The northern mayors also reportedly defied Belgrade by holding a referendum on self-determination on February 14-15, 2012, with over 99 percent of voters rejecting the authority of Kosova's institutions. However, it was unclear whether local leaders were pursuing separate policies or coordinating closely with Belgrade while creating the impression of a rupture so that Serbia would not be blamed for supporting partition.

Belgrade has warned the government in Prishtina against forceful intervention in the north. Internal Affairs Minister Dacic stated in October 2011 that an attack by Kosova security forces on Mitrovica would be considered an attack on Belgrade.[23] Tensions escalated ahead of Serbia's local elections on May 6, 2012, as Belgrade planned to hold them on Kosova's territory. They included a series of arrests and abductions by security forces in both countries in March and April 2012. Kosova's police arrested four Serbs in possession of voting lists and election material, while Belgrade retaliated by arresting two Kosova Albanian police officers patrolling the border.[24]

Kosova Albanians in the Serb-run section of Mitrovica demanded an increased police presence in their neighborhoods after a bomb blast killed an Albanian on April 8, 2012. Since Kosova gained independence in February 2008, Albanians in Serb-majority municipalities have been subjected to various pressures. The opposition Movement for Self-Determination (MSD) urged the President of Kosova to declare a state of emergency in Mitrovica.[25] It also demanded that NATO protect Albanians in the north from violence, while war veterans from the former KLA claimed that if the situation was not resolved, they would "organize the population" to protect itself.[26] Opposition politicians blame the Prishtina government for tolerating the existence of parallel institutions and armed units, thus heightening the likelihood of Albanian retaliation.[27] Serbian gangs prone to violence have formed in the north and are linked to ultra-nationalistic movements in Serbia whose representatives regularly visit the area.

Other developments could also engender conflict. For instance, Prishtina's investment in reconstructing houses for Albanians in the north may spark violence, as Serbs view this as attempts to increase the Albanian presence. At the same time, EULEX is not equipped to respond effectively if violence were to break out. Several countries have withdrawn their special police units that could intervene in controlling crowds. The burden in handling riots and street violence now rests primarily with NATO's KFOR, which is unsuited to such tasks. Moreover, KFOR cannot cover the entire country if violence were to spread beyond the northern municipalities.

During 2012, numerous violent ethnically motivated incidents took place in Kosova. In early April, this included the stoning by Albanian youths of a Serbian delegation attending talks with their Kosovar counterparts in Prishtina, clashes between Albanian and Serbian soccer fans in the Serb enclave of Gracanica, the arrest of four Serbs by Kosovar police for transporting Serbian local election material, and the apprehension of selected Kosovar citizens along the border by Serbian police.[28]

Following intense pressure from Brussels and Washington, Serbia's Minister for Kosova Goran Bogdanovic announced on April 15, 2012, that Belgrade would not organize local elections in Kosova. However, he also asserted that Serbia would not close its institutions in northern Kosova and that instead of elected officials, the Serbian government would appoint municipal leaders.[29] As a precaution against violent clashes during the May balloting, in which Serbs were allowed to vote for Serbia's parliamentary and presidential elections, a KFOR spokesperson requested that NATO send more troops to Kosova. At that time, 550 German and 150 Austrian soldiers were deployed along the lines, dividing the two ethnic communities in Mitrovica.[30]

On June 28, 2012, there were clashes between Kosova police and Serbian activists on the border at Merdare and a subsequent attack on Serb buses in Prishtina.[31] A group of fans from Belgrade Football Club Partizan entered Kosova without the permission of the authorities to attend celebrations marking the 1389 Battle of Kosovo Polje. Kosova President Atifete Jahjaga stated that the men had attacked police with the intention of destabilizing the security situation. Subsequently, Albanian youths attacked Serbs on a bus in Gracanica, a mainly Serb municipality close to Prishtina.

The granting of "special status" for northern Kosova, as demanded by some Serbian leaders, would leave Serbs outside the four municipalities more vulnerable to pressure from the Albanian majority. Kosova's Serbs may perceive such a status as a prelude toward separation regardless of Belgrade's agreements with Prishtina. The achievement of enhanced autonomy would also increase demands for a similar status for the Albanian majority in three municipalities of southern Serbia.

Ivica Dacic, Serbia's incoming Prime Minister at the time, provocatively claimed that the new government would insist on stationing Serbian security forces in Kosova to protect local Serbs.[32] NATO peacekeepers were also placed on alert in preparation for the opening of a Kosova government office in northern Mitrovica. The office would provide services and coordinate investment, while its head, Adrijana Hodzic, would have the authority of a municipal president. Belgrade described the office as a provocation. Oliver Ivanovic, State Secretary in the Serbian Ministry for Kosova, claimed the office would antagonize local Serbs and that the separate local administration would increase its authority in response.

Serbia vs. Bosnia-Herzegovina.

Milorad Dodik, the President of Republika Srpska (RS), one of the two entities in Bosnia-Herzegovina, has periodically asserted that the country will disintegrate.[33] Dodik himself has contributed to promoting such a scenario. In October 2011, he publicly claimed that Bosnia-Herzegovina was a failed international experiment and that the Serbs overwhelmingly supported secession.[34] In response, on October 18, 2011, Bakir Izetbegovic, the Bosniak member of Bosnia's three-member Presidency, published an open letter to Dodik in which he accused him of sowing fear about the future of the state.

Dodik claims that Bosniak Muslims seek political dominance by creating a highly centralized Islamic state while marginalizing the country's Serbs and Croats.[35] Izetbegovic accused Dodik of responsibility for the limbo that Bosnia found itself in by blocking the formation of a central government after the October 2010 elections. Izetbegovic also addressed Dodik's claim that polls showed almost 90 percent of Serbs favoring secession by warning that such statements could precipitate new conflicts. Any attempt to sabotage the territorial integrity of the country would allegedly lead to clashes with "patriots" ready to defend the state. Dodik's representatives subsequently accused the Bosniaks of preparing for war.

Among the potential triggers for conflicts is the status of the Brcko district.[36] In March 2000, Brcko was proclaimed a neutral district by an international tribunal. It was subsequently placed under the authority of Bosnia's central government and run by a multiethnic administration overseen by an international representative. Brcko sits astride a strategic crossroads between the two entities and between Croatia and Serbia. If the RS declares independence, Serbs will want to control the town while the Bosniaks would attempt to sever the entity at various choke points, including Brcko.

Bosniak Muslim and Croat politicians oppose ending international supervision in Brcko, as this could lead to disputes over its neutral status.[37] The Federation parliament accused Bosnian Serb leaders of lacking commitment to respect their obligations toward Brcko. Evidently, the RS government failed to annul its declaration rejecting the international arbitration's decision to declare the district neutral. Bosniak and Croatian leaders prefer that a "national law" be adopted specifying Brcko's representation in central state institutions before international supervision terminates.

Interstate problems also remain between Sarajevo and Belgrade, including incomplete border demarcations, unclear ownership of pre-war assets, the unresolved status of refugees, and untried war crimes cases. Although these are unlikely to precipitate outright conflict, they contribute to the mistrust visible in bilateral relations. Furthermore, the fact that Serbian government officials meet primarily with RS leaders in the entity capital of Banja Luka rather than with Bosnian representatives in Sarajevo also contributes to maintaining tensions.

Serbia vs. Montenegro.

Frictions have persisted between Belgrade and Podgorica since Montenegro voted for independence in a national referendum on May 21, 2006. They have revolved around several grievances, including Montenegro's recognition of Kosova's independence and Serbian government support for nationalist Serbian parties in Montenegro. Podgorica accused the Tadic presidency of direct involvement in incidents designed to undermine Montenegro's independence.[38] For instance, Belgrade supported street protests by Serbian opposition parties and helped them in campaigns to boost the number of people declaring themselves as Serbs in the Montenegrin census of April 2011.[39]

To provide greater support for Serbian populations in neighboring countries, the government in Belgrade has pushed for the creation of a "unified cultural space for Serbs" in the Balkan region and the establishment of a regional board that would implement a national strategy toward all Serbs in former Yugoslavia. Serbia's parliament passed a law on the diaspora and Serbs in the Region in January 2011 that envisages a more intrusive role in nearby states.[40]

The document claims that Serbs are under-represented in public institutions such as central and local governments and that in several post-Yugoslav countries, they are denied human, minority, and religious rights. The document affirms that the Serbian government should play the role of a protector of all Serbs abroad and must provide greater assistance to the SOC for its cultural and educational role among the diaspora and in building a regional network of Serbian organizations.

The text was critically received by governments in Montenegro, Macedonia, and Bosnia-Herzegovina, who viewed it as a prelude to more blatant interference in their domestic affairs. The law states that Belgrade's ultimate objective in Montenegro is for Serbs to achieve "full national rights" as a constitutive and co-equal nation. Belgrade also requested the opening of three consulates in Montenegro, which Podgorica declined on the grounds that these would serve as meeting points for opposition to Montenegrin statehood.

Despite Serbian complaints, the EU Commission declared in October 2011 that Montenegro was implementing its Stabilization and Association Agreement (SAA) in the areas of democracy, the rule of law, and minority rights. The new election law introduced affirmative action for the representation of all minorities in parliamentary elections.[41] In May 2012, the authorities in Podgorica welcomed President Nikolic's statement that Montenegrin independence was not reversible, thereby sending a message to Serbs to reconcile themselves to Montenegrin statehood.[42] However, Nikolic also denied that any differences existed between Serbs and Montenegrins as nations, thereby provoking critical reactions in Podgorica.

Belgrade's backing for the SOC, which claims jurisdiction over Montenegro and does not recognize the Montenegrin Orthodox Church (MOC), also remains a point of contention. Montenegrin officials charge that SOC clergy endeavor to undermine Montenegrin independence. Inter-church battles have revolved around MOC attempts to recover property from the SOC that was seized when the first Yugoslavia was formed in 1918, the MOC was outlawed, and Montenegrin statehood extinguished.[43] The dispute has led to violent clashes between clerics and has been inflamed by statements from Metropolitan Amfilohije, the senior SOC cleric in Montenegro, who remains a firm opponent of Montenegrin independence and has tried to provoke disputes between Montenegrins and Albanians to undermine the new state.

Tensions were raised during the holding of Montenegro's census in April 2011 as Montenegrin and Serb parties launched extensive campaigns to persuade citizens to unambiguously declare their identity. The results demonstrated that the number of Serbs declined slightly and the number of Montenegrins increased since the last census was taken in 2003.[44] According to the 2011 census, ethnic Montenegrins accounted for 44.98 percent of the population, followed by Serbs at 28.73 percent. During the previous decade, the number of Serbs had dropped by 3.26 percent, and the number of Montenegrins rose by 1.86 percent.[45]

Officially recorded demographic trends and the definition of the national language have been major sources of dispute between Montenegrin and Serbian leaders. In October 2007, Montenegrin was declared as the official language of the state, and two new letters were added to the alphabet to distinguish it from Serbian. Nonetheless, 42.88 percent of Montenegrins still consider the language they speak as Serbian, while under 37 percent consider it as Montenegrin. Compared with the 2003 census, people speaking Serbian dropped by 20 percent, and people speaking Montenegrin rose by 14 percent. The language of education also became Montenegrin, and the Latin script has been favored over Cyrillic. In reaction, Serb leaders have demanded that Serbian be declared as the second official language, with the right to a distinct Serbian education also recognized. This could lead to educational boycotts and growing animosity between the two communities that could be exploited by radical political formations.

During 2012, periodic street protests were staged in Podgorica in response to stagnant economic conditions and official corruption.[46] The rallies were attended by nongovernmental organizations (NGOs), students, and labor unions, calling for investigations into all dubious privatizations of former state companies. A protest march on May 15, 2012, ended in front of the government building. Among the demands were an effective campaign against corruption, respect for students' and workers' rights, freedom of speech, and the resignation of the government.[47] Such unrest could assume an ethno-political dimension, pitting Montenegrins against Serbs, if nationalist politicians decided to exploit the protests.

Kosova has been another source of contention between Belgrade and Podgorica. Montenegro recognized Kosova's independence in October 2008, and 2 years later, the two states established diplomatic relations despite Serbia's staunch opposition.[48] Montenegro's President Filip Vujanovic has hesitated in appointing an ambassador to Prishtina until Montenegrins obtain national minority status in Kosova. Kosova's President Jahjaga pledged to meet these demands. According to the 1981 census, the last one recognized by the Kosova authorities, over 27,000 Montenegrins live in Kosova. Out of 120 seats in Kosova's parliament, 20 are guaranteed for Serbs, Roma, Ashkali, Egyptians, Turks, Gorani, and Bosniaks. Montenegrins and Croats are to be included on this list.[49]

Vojvodina Spotlight.

Vojvodina is a multiethnic region in northern Serbia in which autonomist sentiments are held across the ethnic spectrum. According to the 2002 Serbian census, from a population of two million, Serbs form 65 percent and Hungarians 14.28 percent, with the remainder divided between over a dozen ethnic groups.[50] Although separatist conflicts are not imminent in Vojvodina, frustration with Belgrade and the lack of investment in the region's economy are propelling sentiments toward greater self-determination and even statehood. In opinion polls taken at the close of 2011, nearly 10 percent of inhabitants favored Vojvodina becoming a republic, with extensive autonomy within Serbia or even a separate state.[51] In earlier surveys, the number of supporters of the province's independence only ranged between 1 and 2 percent.

On November 19, 2009, the Vojvodina Assembly declared a new statute for the province, which was accepted by the Serbian parliament. The statute has been criticized by Vojvodinian autonomists as well as by Serbian nationalists. The autonomists argue that it fails to strengthen the province's executive, legislative, and judicial powers and should be replaced by a separate Vojvodinian constitution.[52] They complain that Belgrade's failure to pass a law on funding Vojvodina has negatively affected economic development and that the province should control all of its resources. The President of the Vojvodina Assembly, Sandor Egereši, contends that Vojvodina's constitutional jurisdictions need to be resolved by defining the model of financing and allowing Vojvodina to form its own police force.[53] Vojvodina's Prime Minister Bojan Pajtić has stated that the territory acquired only two-thirds of its competencies in the process of transferring authority from the state to the province, as stipulated by the new statute.[54]

Serbian nationalists expressed outrage over Vojvodina's new autonomy statute, viewing it as the thin end of a separatist wedge leading to independence.[55] In particular, the Progressive Party, the Socialist Party, and the Serbian Democratic Party, who favor Serbia's centralization, contend that Vojvodina is becoming a quasi-state that will lead to the further disintegration of the country. In protest against growing autonomy, about 12 of 50 towns and municipalities in the province refused to hoist the flag of Vojvodina on their town halls, despite the stipulations of the regional government.[56]

In another controversy, Serbian nationalists protested against the inauguration of a Brussels office for Vojvodina on October 10, 2011, alleging that the region's leaders were implementing a separatist agenda.[57] Vojvodina officials argued that the office would enable the region to access European funds and foreign investment. In fact, over 300 regions from various European states maintain offices in Brussels without provoking accusations of separatism.

On July 10, 2012, Serbia's Constitutional Court ruled that a law granting Vojvodina autonomy was unconstitutional.[58] This ruling ensures that the province will no longer have jurisdiction over environmental, agricultural, and rural development policies. Nationalist political parties welcomed the court's decision. Vojislav Kostunica, leader of the Democratic Party of Serbia, claimed the court had defended Serbia from further separatism. The court also disputed the designation of Novi Sad as the capital of Vojvodina and the existence of Vojvodina's offices in Brussels.

Belgrade's actions may have the reverse effect of the one intended by mobilizing activists in Vojvodina to demand more far-reaching self-determination and thereby engendering new conflicts with the authorities in Belgrade.

III. BOSNIAK NATIONAL REVIVAL

Seventeen years after the end of the 1992-95 war, Bosnia-Herzegovina continues to struggle with inter-ethnic reconciliation and integrated statehood. The country is divided along ethno-national lines, with citizens polarized around their respective identities.[59] The dysfunctionality of the Bosnian state, coupled with Serbian support for full administrative autonomy or even secession and Croatian dissatisfaction with minority status, has spurred Bosniak nationalism as a defense against rival political and territorial claims.

Moves toward Serbian separatism can both nourish secular nationalism and raise religious identification among Bosniak Muslims, which will breed radicalism and provoke conflict. Bosnia's institutional gridlock and economic stagnation has resulted in growing disillusionment with multiethnicity or triethnicity, which blocks decisionmaking at various administrative levels. Meanwhile, nationalist leaders perceive the civic option, or the principle of citizenship not based on ethnic identity, as harmful to the group interests of all three major collectivities that they claim to represent.

Stalled State Building.

The state-building project in Bosnia-Herzegovina has stalled. The country has entered an era of uncertainty and faces increasing threats to its stability and territorial integrity.[60] The prevailing assumption by EU leaders that the scaling down of international supervision and the magnetic attraction of EU integration would convince Bosnia's political leaders to pursue the rigorous reforms necessary for accession is proving illusory. On the contrary, in recent years, Bosnia has regressed as a functioning state and is increasingly riven by disputes between leaders of the three major ethno-national groups. The country faces drift and division that could culminate in destructive new conflicts that would impact negatively on the wider region.

The Dayton Accords, signed under international supervision in November 1995, were instrumental in terminating armed conflicts and constructing an administrative structure that pacified the three major ethno-national groups. One of the primary goals of Dayton was to give the three nations a stake in remaining in a single country through a protective veto over decisionmaking. However, the agreement was not designed to build an integral and effective state in which the central government in Sarajevo possessed decisive authority. Instead, it created a complex administrative structure in which ethnic balancing predominated and layers of governmental bureaucracy contributed to inefficiency and budgetary burdens. This system has obstructed decisionmaking where ethno-national interests prevail over civic-state interests.

The Dayton process was overseen by the Peace Implementation Council (PiC) Steering Board, mandated by international institutions. The main instrument for pushing through reforms and upholding the unity of the state has been the Office of the High Representative (OHR). When the OHR used the Bonn Powers, approved by the PiC in December 1997 to enable more intrusive and effective intervention, Bosnia consolidated its statehood. When the OHR took a back seat and urged dialogue and compromise between nationalist leaders, little progress was achieved.

However, the prominent OHR presence has also provoked charges of international interference, which fosters a dependency relationship and ignores the democratic choices of Bosnian citizens. Indeed, international actors confront a major dilemma regarding means and ends: whether it is preferable to impose a more centralized state that can make progress toward EU and NATO membership or if it is more empowering and democratic to permit local leaders to obstruct the process and pursue essentially separate ethnonational agendas.

A limited and weak international role emboldens Bosnian leaders to conclude that certain rules in pursuit of statehood will no longer be enforced. The major nationalist parties are more interested in preserving their particularistic interests than in constructing an integral state. This was evident after the October 3, 2010, general elections when Serbian and Croatian national parties blocked the creation of a new state-level government by the election winners, the multiethnic Social Democratic Party (SDP), until the end of December 2011.

Attempts at constitutional reform to prevent entity and ethnic blocking of state legislation and ensure smoother government operations have been obstructed through entity voting. National leaders rejected the April package of constitutional reforms proposed by international mediators in 2006 and designed to make the government more efficient. Renewed U.S. and EU efforts for constitutional reform during the October 2009 Butmir Summit also failed to bring results.

The debate between policymakers favoring international pressure on local leaders to push through reforms and those who believe that nothing durable will be achieved unless local leaders voluntarily agree without foreign interference is a constant factor in Bosnian politics. Proponents of a strong OHR presence contend that without internally generated progress, the international office must be maintained to prevent regression and damaging political conflicts. Indeed, Western officials and analysts concerned about Bosnia's survival favor a strong OHR or its replacement by an equally effective EU High Representative.

The most destabilizing outcome would be the disappearance of the OHR and a weak and divided EU mission that is unable to discourage Bosnia's political fractures. The OHR is seen by Bosniaks in particular as a safeguard that ensures close U.S. involvement. There are fears that the EU delegation in Sarajevo aims to remove the Americans without an effective replacement. This may suit national leaders who discount any mediating role for outside powers and believe that, without international interference, Bosnia-Herzegovina will move toward a confederal arrangement or an outright territorial division.

Ethnic and Civic Politics.

The EU is rhetorically supportive of Bosnia's multiethnicity, but it has been critical of the multiethnic SDP since the October 2010 elections, claiming that it could destabilize the inter-ethnic or internationalist political arrangements and prove troublesome for EU monitoring. Paradoxically, the EU endeavors to maintain the status quo and uphold relative political stability even if this disables progress on the reforms necessary for Bosnia to enter the Union.

Ethno-politics has dominated Bosnia's governance since the end of the 1992-95 war and has stymied the development of state citizenship, programmatic pluralism, individual rights, and a competitive democracy. Ethno-nationalist parties, treated as the sole representatives of ethnic collectivities by international actors, are primarily based on patronage and clientelist networks, and their leaders are adamant that the civic principle cannot be applied in Bosnia but only a system of intergroup balancing can.

In October 2010, for the first time in 15 years, a civic based party, the SDP, won the national elections. This indicated a growing constituency for a nonethnic vote that undermines the ethnic party stranglehold on political institutions. Serb and Croat nationalist leaders claim that the civic project is primarily a cover for pursuing centralization, promoting Bosniak domination, and ensuring the minoritization of the Serb and Croat populations. Some analysts even contend that the SDP's focus on civic identity is provoking Serb and Croat nationalist leaders to push for separation. Paradoxically, the more successful the civic project becomes, even though it is based on EU norms, the more it contradicts the principles of Dayton, particularly the legitimacy of ethnic balancing and entity blocking, and can precipitate national conflicts.

There is no single Bosnian political elite that transcends national divisions, and no common pan-Bosnian identity has emerged since the war.[61] There is also an absence of practical unification projects, such as infrastructure construction, educational reform, or common youth movements that would help bond the three nations within a single state. Bosnian state identity remains shallow and artificial. The younger generation has no tradition of multiethnic Yugoslavism, and there is minimal interaction between ethno-national groups. For instance, Serbs from Banja Luka are more likely to visit Belgrade or Zagreb than Sarajevo. The educational systems are separated, and there is no daily interaction between citizens in the two entities. This leaves young people susceptible to indoctrination and political manipulation.

The new government of the Federation of Bosnia-Herzegovina (the joint Bosniak and Croat entity) was constituted in March 2011 but not recognized as legitimate by the RS National Assembly or by the major Croatian parties, the Croatian Democratic Union (HDZ), and the Croation Democratic Union (HDZ-1990), the later having split from the former in April 2006. The HDZ-1990 protested their exclusion from the cabinet although Bosnian Croats from other parties were represented in the administration. The two nationalist parties, which captured almost 80 percent of the Croatian vote, predominantly in western Herzegovina, asserted their political monopoly by arguing that only they could nominate genuine Croatian representatives and demanded all five ministries assigned to Croat delegates. Nationalists do not consider Bosnian Croat politicians elected by members of other ethnic groups as authentic national representatives. By contrast, the civic-focused SDP argues that lawfully elected Croats, regardless of who voted for them, can represent the community.

Croats have felt increasingly marginalized in the Federation, which contains a larger Bosniak population, while more Croats have left the country. Croatian numbers in Bosnia-Herzegovina as a whole have decreased from 17.38 percent of the population in 1991 to 15.4 percent in 2000.[62] The Bosniak proportion rose from 46 percent in 1991 to 48.3 percent in 2000, and the Serbian population also increased from 31.21 percent in 1991 to 37.1 percent in 2000, although largely because of the declining Croatian proportion. In 2012, the estimated population of the Federation stood at 2.5 million and the RS at 1.2 million. The Bosniak proportion in the state as a whole is believed to number over 50 percent, with Serbs forming under 40 percent. Croatian numbers have shrunk to approximately 10 percent, with more inhabitants expected to leave Bosnia when Croatia enters the EU in 2013.

Serbian leaders claim that their population would also significantly diminish if they did not have the RS to protect their interests. Indeed, a falling Serbian population could make the government in Banja Luka more eager to push for secession to preclude the prospect of a diminishing political role. Changing proportions have also led to disputes over holding a new census, as Croatian leaders in particular want to maintain the power sharing arrangements based on the 1991 census when their population was larger.

To protect their collective interests, some Croatian leaders in the HDZ have voiced support for the creation of a third entity, a Croatian majority unit, to be carved out of the Federation. Such a scenario would also result in the establishment of a Bosniak Muslim entity and enable the RS and a potential Herzeg-Bosnia entity to act in unison to obstruct the state government in Sarajevo. The RS leadership is supportive of such restructuring and is reportedly coordinating its approach with the HDZ to defend the politics of ethnic blocks and nationally-based economic interest groups.

The RS, similarly to the Federation, was also established as a multiethnic entity where no major governmental decisions were to be taken without agreement between representatives of all three nations. In practice, the Serbian parties remain dominant, and the RS President makes all essential decisions without consultation with Bosniak or Croatian leaders. Indeed, the RS is evolving into a one-party quasi-state. In stark contrast, decisionmaking in the Federation requires the consent of representatives of all three national groups.

Role of Radical Islam.

Serb and Croat nationalist leaders claim that the Bosniaks are becoming radicalized through militant Islam. By alleging that Wahhabism and Salafism, ultra-conservative streams in Islam, are growing among Bosniaks, RS leaders can pose as defenders of allegedly endangered Serbian national interests. In reality, Islamist political influence is not a mainstream phenomenon, as the overwhelming majority of Bosniak Muslims belong to the moderate Hanafi school of Sunni Islam.[63] Their secular attitudes have sparked disputes with foreign Islamic radicals seeking to proselytize their puritanical beliefs.[64] Nonetheless, as in other European countries, radical Salafi streams do exist outside the control of the official Bosnian Islamic Community, and some of their members may be prone to a violent jihadist ideology.

A Bosniak national identity with religious identification has developed since the 1992-95 war and is focused on maintaining the integrity of the Bosnian state. Islam has served as a tool for building an ethnic identity rather than being the "final destination of identity politics."[65] The notion of an Islamic Bosniak state does not attract young people. It is estimated that less than 10 percent of the Bosniak population favor partition and the creation of a Muslim Bosniak republic.

The growth of Muslim nationalism and Islamist influence is more likely if the country starts to splinter. This would be a response to Serbian and Croatian separatism that could intensify the struggle within the Islamic Community over the future of a smaller Bosniak state. A partitioned Bosnia would heighten the grievances felt by the chief victims of the war and would convince a growing number of Bosniaks that they had been betrayed by the Western powers. It could also open up new ground for radical religious influences.

The Wahhabi movement in Bosnia-Herzegovina is not native but was "imported" during the 1992-95 war when mujahideen fighters from different parts of the world volunteered to defend the Muslim Bosniaks. Their maximum number was estimated at about 4,000 by the close of the war, and they generally lived in isolated rural communities. After the signing of the Dayton Accords, about 1,300 remained in the country and acquired Bosnian citizenship. Many of these subsequently lost their citizenship under pressure from the government, and only 200 were left in the country by 2010, most of them married to local Bosniak women.

The terrorist Mevlid Jasarevic who fired shots at the U.S. embassy in Sarajevo before being apprehended on October 28, 2011, adhered to the radical Takfir ideology advocating intolerance toward non-Muslims and defying secular laws. He reportedly communicated with Wahhabis in the village of Gornja Maoca near Brcko.[66] Members of this group were believed to have planted an explosive device outside Bugojno police station in June 2010 that killed one police officer. Ismet Dahic, former head of the police in Sarajevo, claimed it was possible that Serbian police agencies recruited

Mevlid Jasarevic and sent him to Sarajevo to discredit the Bosnian state.

The main recruitment center for Balkan Wahhabis is believed to be in Vienna, while the major outside Muslim influences in Bosnia emanate from Turkey and other moderate Islamic states. Turkey has increased its economic and cultural influences among Islamic populations in the Balkans, but it does not exert a decisive role in Bosnian politics despite Ankara's aspirations as a regional leader. Nonetheless, Turkey does contribute to undercutting Salafi influences among Muslim populations.

Ankara has also posed as a regional mediator with varying degrees of success.[67] For instance, Ankara mediated a dispute between rival Bosniak groups in the Sandzak in southern Serbia, although the deal subsequently collapsed. It has also claimed success in reconciliation between Belgrade and Sarajevo through a Trilateral Consultation Mechanism that led to the appointment of a Bosnian ambassador to Belgrade and the signing of the Istanbul Declaration that "guaranteed the territorial integrity and sovereignty of Bosnia-Herzegovina."[68] However, Turkey's alleged indispensability in this process has been disputed. Moreover, Balkan leaders avoid creating the impression that they are moving closer to Turkey and surrendering their EU aspirations.[69]

Separatist Maneuvers.

For Bosnian Serbs, the prospect of EU membership is becoming less convincing than the ambition of independence. The optimum objective for leaders of the Serbian entity in Bosnia-Herzegovina is statehood. To achieve such a goal, a three-pronged strategy has been pursued: preventing the creation of an authoritative central government in Sarajevo, fanning disputes between Bosniak and Croat politicians by supporting greater Croatian autonomy, and gaining prestige and power through political brinkmanship such as preparations for public referenda in the Serb entity.

RS President Milorad Dodik has persistently claimed that the Bosnian state is not functioning, and the RS needs to develop its sovereignty. He has opposed the state government on the grounds that it is abrogating powers that belong to the entities and is engaged in a policy of centralization and Bosniak Muslim domination.[70] Dodik says he is reclaiming as much autonomy for the Serb entity as possible and asserts that the state government in Sarajevo should only deal with foreign and security policy.

In public opinion surveys, majorities of Croats and Bosniaks oppose the potential secession of RS, whereas 87 percent of respondents in the Serb entity support the creation of an independent state if a majority of its citizens voted for it.[71] These surveys revealed that 56 percent of Croats and 86 percent of Bosniaks did not agree with the idea of dividing Bosnia, while 61 percent of Serbs support such a plan. Additionally, while 62 percent of Serbs identified strongly with their entity, as did 46 percent of Bosniaks, only 11 percent of Bosnian Croats empathized strongly with the Federation. The Bosniaks were the only group in which a significant percentage of respondents (44 percent) identified with the country of Bosnia-Herzegovina.

Dodik claims that he does not seek RS secession but simply the return to entity level of governing prerogatives captured by Sarajevo in recent years and the prevention of any further erosion of entity powers such as the elimination of the "entity veto" in the passage of Bosnian legislation. Dodik contends that he seeks a confederation between three sub-state entities with a weak central government but has also indicated that Montenegro's referendum on independence in May 2006 may serve as a model for the RS. Although Dodik began as an anti-nationalist, he has adopted a more ethnocentric approach in support of RS separation. Some analysts think that Dodik sees himself as the unifier of Serbian lands and may have ambitions to preside over a joint state with Serbia.

Dodik benefits from provoking confrontation with Sarajevo and courting direct talks with EU officials and the EU Commission. This was the case during the involvement of EU High Representative Catherine Ashton in May 2011, who appealed directly to Dodik to suspend a planned RS referendum on the State Court, which leaders in Banja Luka claim is biased against Serbs. They wanted to repeal a provision that allows the state-level court to take over entity cases. The visit of Ashton to Banja Luka raised Dodik's stature as a statesman and leader of an aspiring state.

Dodik has continued to dangle the specter of referenda on such issues as OHR legitimacy, state property, judicial reform, and potentially the RS abandonment of state institutions in pursuit of a confederation of two independent states. He remains patient and opportunistic, tests international resolve, and retreats from some controversial step when there is a firm international response but pushes ahead when the reaction is tepid.[72] In the longer-term, RS leaders calculate that international interest in Bosnia will wane, the United States and NATO will become disengaged, and the EU will remain divided. This will allow the RS to move toward secession and unification with Serbia, thus enabling Dodik to leave a lasting legacy in Serbian history.[73] In the meantime, he has created a centralized and authoritarian RS in which the Alliance of Independent Social Democrats controls the government, presidency, and National Assembly and in which Bosniaks and Croats play a secondary role.

The RS works more effectively as a state than the Federation, as it is a more streamlined structure without cantonal level administrations. In justifying its opposition to state institutions, Banja Luka complains about the state government in Sarajevo as mismanaged, inflated, wasteful, and corrupt. RS leaders assert that over 90 percent of their population favor separation and claim that Bosnia is a failed state that was incapable of forming a durable government.[74] In response, Bosniak leaders charge that the RS was created through genocide and ethnic expulsions, and it has become increasingly autocratic and centralized under Dodik, while Bosniaks and Croats are excluded from decisionmaking.

Dodik also seeks to leverage the EU to neutralize the OHR and the U.S. presence. Nonetheless, he has also needed the OHR to claim he is defending Serbian interests against unwarranted international interference. His evident preference would be for a weak and temporary OHR before he makes a final decision on separation. His game of brinkmanship with Sarajevo and international players has thus far proved successful, but there is always a possibility of miscalculation that may provoke a violent reaction.

Some Croatian activists angered by the exclusion of HDZ and HDZ 1990 from the federal entity government have focused on creating a regional intercantonal structure on 20 percent of Federation territory, styled as the Croatian National Assembly (CNA) and based in Mostar. The CNA is intended to coordinate municipal and cantonal administrations in which Croats form majorities and some of its leaders have demanded the formation of a third Bosnia entity. Their moves are openly supported by Banja Luka. More radical Croatian activists have either called for a distinct Croatian entity or the re-establishment of the wartime Republic of Herzeg-Bosnia.

State institutions continue to be obstructed by distinct ethno-political interests at entity level and register minimal progress in meeting EU stipulations necessary to attain candidate status. For instance, Croat and Serb leaders have opposed Sarajevo's efforts to centralize decisionmaking by transferring police authority from the cantonal to the entity level. Even at municipal level, conflicts over inter-ethnic power sharing persist. In the starkest example, Bosniak-Croat relations remain tense over the divided city of Mostar.[75] The core of the dispute revolves around two Constitutional Court rulings, one in June 2011 and one in January 2012, that the election of three delegates from each of the six city areas to the City Council was unconstitutional. The rulings assert that it is unconstitutional that an area of the city with a population of almost 30,000 Croats elects the same number of delegates to the town assembly as an area with 7,000 Bosniaks.

Croat leaders propose that Mostar should be divided again into municipalities, leaving the City Council to deal only with joint issues.[76] The HDZ contends that when the former High Representative Paddy Ashdown abolished Mostar's municipalities in 2004, his objective was to disable Croats from having more power than Bosniaks, thus discriminating against Croatian residents. Among Bosniaks, the Constitutional Court rulings are perceived as a direct threat to their existence in Herzegovina and would allegedly lead to Mostar becoming a capital for the Croatian dominated cantons. Local Croats view opposition to the recent ruling as confirmation that Bosniak leaders are seeking to turn Croats into a minority with lessened rights. Most Croats in Herzegovina believe that "losing" Mostar would make life untenable for the community throughout Bosnia. They are equally concerned that Bosniak leaders are seeking to centralize the Federation.

One additional factor encouraging partition has been the ongoing division of state property, including land and utilities, and the legalization of its ownership by the two entities. The RS is preparing a new property law that would effectively reduce Bosnian state holdings. Such a process would make Bosnian statehood increasingly tenuous.

Economic conditions have been stagnant in both entities for several years. Bosnia's budget bears a heavy burden with the enormous state sector acquired through the Dayton Accords. Although the five administrative levels (city, municipal, cantonal, entity, and state) and the Brcko district government provide thousands of civil service jobs, they also inhibit state investment in productive business. Plans to trim the state sector and lower its budget will depend on entity agreement. Although several cantons are financially unsustainable, the HDZ will claim discrimination if the cantons are merged or eliminated and if the Federation government is strengthened at the expense of the cantons, as some politicians in Sarajevo have proposed.

Bosnia's state structures depend largely on outside funds. Any prolonged absence of a state government creates problems in releasing EU resources and tranches of IMF loans, and it curtails prospects for foreign investment. Some observers believe that financial restraints in the RS finally convinced Dodik to allow for the creation of a state government in December 2011 to fill growing gaps in the RS budget from EU and IMF sources. However, Dodik has also been courting Russia, Serbia, and China to access alternative funding sources and has sought to attract foreign investors for the RS, while bypassing Sarajevo.

Regional Reflexes.

Serbia's President Boris Tadic walked a tightrope between satisfying EU demands that he recognize Bosnia-Herzegovina's integrity and Serbian nationalist pressures that he support RS President Milorad Dodik. Dodik's high popularity in Serbia meant that Tadic could not be seen to oppose him. Indeed, Belgrade intensified its relations with Banja Luka through such mechanisms as joint intergovernmental sessions, while Tadic rarely visited Sarajevo during his frequent trips to Banja Luka. Although he affirmed Bosnia's independence, Serbia's newly appointed Prime Minister Ivica Dacic has publicly stated that the RS and northern Kosova should be joined with Serbia. This has given ammunition to Bosniaks who charge that the Greater Serbia project has not been abandoned by Belgrade. Indeed, some Serbian politicians view the RS as compensation for the loss of most of Kosova.

The Croatian government is generally supportive of the HDZ in an effort to protect Bosnian Croat interests, and Zagreb was dismayed by the HDZ's exclusion from the Federation government after the October 2010 elections. On the other hand, Croatia has sought a closer relationship with the government in Sarajevo and does not appear to have a parallel agenda of partition. By contrast, Belgrade remains under suspicion in

Sarajevo that its pursuit of a special relationship with the RS is a prelude to supporting secession.

The prospect of EU accession alone is not magnetic enough to stimulate reform of the state, especially as the gap between promise and reality appears to be widening. Although over 80 percent of the public supports EU membership, the slow process of entry obstructed by political and structural factors inside Bosnia fuels public frustration and susceptibility to ethno-nationalist appeals. In terms of its EU prospects, Bosnia-Herzegovina suffers from three core disadvantages: It is a disunited state, an unreformed state, and an unwanted state. Sarajevo's bid for accession will be slowed down by its divided polity, collectivist mentality, and paternalistic state structure, as well as by potential EU rejection of a semi-Muslim country.

The Dayton Accords do not meet the criteria for EU entry, and it remains unclear whether the OHR needs to be closed before the EU would consider Bosnia a credible candidate for membership. There is an inbuilt paradox in this equation: A strong OHR, European Union High Representative (EUHR), or head of the EU Delegation is evidently needed to push through EU conditions for accession. However, EU states are unwilling to use the Bonn Powers or any other instruments to achieve such an outcome, arguing that this would mean international imposition. Moreover, the EU has leaked credibility, as it moved slowly in establishing its mission in Sarajevo and only appointed Peter Sorensen as head of the EU Delegation at the end of May 2011.

The EU set three conditions for Sarajevo to gain access to pre-accession assistance funds and before it could submit a formal application for membership. Bosnia's progress toward EU accession has been delayed because of failure to meet these conditions. First, the government needed to adopt a state aid law to control the level of state subsidies to the public and private sectors. Second, Sarajevo had to approve a census law at state level to provide the legal basis for the first national census since 1990.

Third, the government needed to reform the state-level constitution and election laws to comply with the 2009 ruling of the European Court of Human Rights (ECHR) and terminate ethnic discrimination against representatives of minority groups, such as Jews or Roma. The latter are excluded from holding government positions as they are not members of the three state-forming nations. Disputes raged during the summer of 2012 over reform of the election laws, as leaders of the three state-forming nations resisted changes that would dilute their power by allowing for the representation of other national groups as well as people who did not identify with any single ethnicity.

Political developments in Bosnia will be determined by the aspirations of local leaders, the reactions of ethnic counterparts, and the stance of international actors. Ultimately, Bosnia faces one of two scenarios—the optimistic and the pessimistic. In the optimistic version, Bosnia experiences more rapid progress toward NATO membership, as EU access is a much longer-term proposition. NATO can provide an overall security umbrella, discourage separatism, consolidate Bosnian borders, promote civil-military reform, help modernize the armed forces, and give a reformist boost toward EU accession.

Bosnia has obtained Membership Action Plan (MAP) status with NATO, and its Alliance entry has been supported by leaders in both the Federation and the RS, although growing Russian influence may temper such sentiments in Banja Luka. While the majority of RS residents are reportedly opposed to NATO entry, a majority in the Federation remains supportive. Progress toward NATO would be promoted through a settlement over state property, including military facilities that are currently claimed by entity governments. A divided military, where only the officer corps is integrated and disputes persist over property, cannot be a credible NATO candidate.

Dangers of Bosnia's Division.

In the pessimistic scenario, Bosnia-Herzegovina slides toward open conflict and violence. The deterrence capabilities of the EU may be insufficient to stymie armed clashes, with European Forces (EUFOR) having fewer than 1,500 troops in the country. They are poorly prepared to respond to outbreaks of violence and are likely to downsize further. EUFOR's crowd control and counterriot capabilities are reportedly inadequate, and reducing numbers sends the wrong signal at a time when tensions can escalate.

Even if a new war does not materialize, Bosnia could become increasingly ungovernable if ethnic disputes intensify and the legitimacy of the state government is further eroded. Numerous grievances remain in Bosnian society, which hinder inter-ethnic reconciliation. These include restricted refugee returns, the limited recovery of property confiscated during the 1992-95 war, and lack of sufficient compensation by the RS authorities to the Islamic Community for destroyed mosques.[77]

Potential provocations could lead to violence, such as the RS holding a referendum on sovereignty or withdrawing Serbian representatives from state institutions, an attempt to seize Brcko, the desecration of a mosque, or the creation of checkpoints along the inter-entity boundary line. RS officials plan to finalize inter-entity border demarcations, while political leaders in the Federation remain reluctant. Representatives of the largest Bosniak nationalist formation, the Party of Democratic Action (SDA), claim that RS officials are intent on presenting inter-entity lines as permanent borders.[78]

Conflicts can also erupt between Bosniak and Croat activists in the divided city of Mostar, especially if Croatian national parties push for a third entity or greater decentralization of the state or if the Federation government decides to reduce the number of cantonal administrations. In an indication of brewing tensions, riots broke out in Mostar on June 19, 2012, in the wake of the defeat of Croatia's football team at the Euro 2012 championships.[79] Hooligans from the western, Croatian part of town clashed with police as they tried to reach the eastern, or Bosniak, section. Violent incidents were reported near a boulevard that marked the wartime separation line, but police prevented the crowd from reaching the eastern part of the city.

In a survey of youth attitudes conducted in early 2012, only a few respondents thought that the root cause of conflict might be inter-ethnic tensions. Nonetheless, many believed that once started a conflict would play out along ethnic lines.[80] Respondents felt that violence between individuals was not caused by ethnic differences but could acquire an ethnic dimension. Although the majority of citizens oppose a new war, a small minority may favor armed conflict in an environment where people are easily manipulated, partly for fear of becoming victims themselves. Some respondents felt that years of political crisis have created an environment of fear, and it would take a small spark to start a new collision. Many people listed the poor condition of the economy as a potential trigger for violence, as economic frustration can spill over into inter-ethnic confrontations.

A local firefight could spark a broader conflagration. Bosniaks may be prepared to fight to keep the country intact, as the idea of peaceful separation is widely dismissed as illusory and opposed by the vast majority. Bosniaks would perceive the breakup of the country as a delayed defeat from the 1992-95 war in which genocide would be legalized and the Greater Serbia and Greater Croatia projects given credence. The main reason the Bosniaks signed on to Dayton was that it ensured state continuity in the post-war setting. If the partition option was pursued, in addition to Bosniak nationalism, we could also witness growing militancy among supporters of the civic option and a reaction against national divisiveness.

Some local analysts and political leaders estimate that it would not be difficult to mobilize 100,000 Bosniak volunteers to fight against the RS.[81] Sarajevo's strategy would be to sever the entity at several jugular points, particularly at the Brcko intersection in northeast Bosnia and across Srebrenica or Gorazde toward the Drina River. This could result in new rounds of ethnic expulsion and altered dividing lines that would favor the Bosniaks, as Serbs would be unable to defend their entire territory. Banja Luka would likely appeal to Belgrade for military assistance and raise the danger of interstate clashes.

In such a scenario, nationalist Bosniak and civic leaders could announce the abandonment of the Dayton two-entity system to strengthen the central government in Sarajevo. Bosnian Croats may respond by creating their own separate government in western Herzegovina and appeal to Croatia for support and armed volunteers. There is also a possibility that Sarajevo and Zagreb may coordinate a military response against attempts at RS secession, especially if Banja Luka is backed by Belgrade.

For several years, Bosnia-Herzegovina has been ranked among the most vulnerable states in terms of risks for internal disturbance. In October 2011, a report released by the U.S.-based Democratization Policy Council and the Sarajevo-based Atlantic Initiative issued warnings about renewed inter-ethnic violence.[82] The authors did not predict imminent conflict, but their observations need to be heeded by international actors who assume that the current status quo can last indefinitely. They criticize the role of international agencies in their unwillingness to effectively employ Dayton enforcement mechanisms such as OHR and EUFOR. The "soft power" approach has enabled local political leaders to pursue their agendas unrestrained, and there is no collective political will to resist. The PiC Steering Board is divided between members believing EU membership prospects will be sufficient to prevent further deterioration and capitals such as Washington and London who are increasingly frustrated by this approach. International disunity stimulates radical Bosnian agendas.

The Atlantic Initiative report pinpoints a number of factors that encourage extremism and conflict, including inflammatory political rhetoric, dysfunctional state institutions, and the impact of the global economic crisis. Political discourse in Bosnia has deteriorated sharply since the October 2010 elections, with more politicians questioning the existence of the Bosnian state and using the 1992-95 war as their main reference point. In such a climate, popular fears appear to be rising. Confrontational relations between political leaders and the deteriorating performance of governing institutions at every level have also contributed to the malaise. Political leaders have been unwilling to compromise on policies that serve the public good, and the basic needs of citizens are largely unmet. This raises public anger toward the state and a sense of growing insecurity.

The global economic downturn has also affected Bosnia, with GDP growth turning negative in 2009 and a tepid recovery in 2010 and 2011. Unemployment in Bosnia is one of the highest in Europe. Official numbers indicate 42 percent, although the figure is closer to 30 percent if account is taken of unregistered employment. The onset of the recession and reductions in remittances from Bosnians working abroad have exacerbated budgetary shortfalls, especially at the entity level, and will impoverish more households and raise the likelihood of social protests.

The cumbersome state structure, with five levels of administration, remains a major burden on the state budget. The country is also bedeviled by systemic corruption, an unreformed judiciary, and a patronage network that drains the economy to a point of potential insolvency. Economic deterioration, the lack of opportunities, frustration with pervasive nepotism and corruption, and rising social tensions can ultimately lead to violent social unrest that assumes ethnic dimensions.

There are several potential triggers for violence, whether organized or spontaneous. Football hooliganism and juvenile delinquency imbibed with nationalist indoctrination in the midst of a faltering economy can trigger intergroup conflict. Hooligan gangs and criminal networks can be manipulated for political goals in an organized fashion, as they are relatively easy to mobilize. Private security firms with professional personnel and an easy availability of weapons can also be deployed if organized violence were to erupt.

Minority returnees remain a vulnerable social group, as relations with the ethnic majority in their neighborhood remain strained, and their access to jobs, housing, and social services is restricted. The increasingly heated political environment has added to their feeling of insecurity, while incidents of violence have the potential to snowball into wider interethnic clashes.

Allegations about a growing terrorist threat in Bosnia, based on the claim that the number of Wahhabis and Salafis is increasing, are not aimed at deterring such a threat but at branding Bosniaks as terrorists and delegitimizing their political aims. In response, the official Islamic Community has denied any security threat posed by a few dangerous individuals and dismissed all such references as Islamophobic. Such hardened positions foster polarization and mutual prejudice.

Although Bosnia has ranked for many years among countries with the lowest recorded number of terrorism-related incidents, it does have several hundred ultra-conservative Salafis among whom there may be potential terrorists.[83] Although these individuals currently pose a limited danger and have marginal public support, they could exploit opportunities created by a deteriorating political climate to pursue their agendas. At the same time, effective deterrents remain weak, borders are porous, and an abundance of readily available weaponry exists.

In an indication that religious radicals may have an increasing influence on some aspects of government policy, Sarajevo canton's education minister, Emir Suljagic, resigned in February 2012 after repeated threats to his safety from hardline Muslims. The latter opposed educational reforms that purportedly downplayed the importance of religion and upheld secularism.[84] Some analysts believe that political clericalism among all three ethnic groups is becoming an increasingly aggressive force.[85]

The Atlantic Initiative report also casts doubts on the effectiveness of law enforcement. The police are subjected to political pressure to submit to ethnic political loyalties. Police capacity to uphold public order in the event of violent inter-ethnic incidents remains in question, and the force would likely split along ethnic lines in the event of any large-scale conflict. The professionalization of the military has also been stunted by political infighting and disputes over the ownership of defense property between entity and state administrations. The military has downsized to under 10,000 professional soldiers. Its ethnic-based infantry battalions could become embroiled in civil conflict, together with the country's veterans' organizations and private security firms.

The abundance of weapons remaining from the war and the relative ease with which they can be obtained is especially dangerous.[86] Bosnia's Ministry of Defense estimates that it stores around 95,000 surplus weapons and some 25,000 tons of surplus ammunition. Arms stockpiles under official supervision are often poorly guarded. In addition, large amounts of weapons remain outside government control. Many of these arms, mostly AK-47 assault rifles, rocket-propelled grenades, handguns, and hand grenades, were stashed illegally out of a fear of renewed conflict.

Reportedly, every third citizen owns arms, and every fifth citizen possesses an illegal firearm.[87]

Despite these negative trends, the EUFOR contingent is not an effective force for concerted deterrence. It has shrunk from 7,000 troops at its launch in December 2004 to an estimated 1,300 by August 2011. The downsizing of the force has left it without forward bases outside Sarajevo. It lacks helicopter lift for operational purposes, and unilateral withdrawals by several countries means that the force is incapable of fulfilling its obligations. In sum, EUFOR has lost the ability to provide a credible deterrent and may fail to maintain or restore security, especially if hostilities were to erupt in several Bosnian towns simultaneously. Instead, militants may exploit a growing security vacuum, and the KFOR contingent in Kosova may not be in a position to reinforce EUFOR, as it has also been downsized and faces its own simmering security threats.

The Atlantic Initiative report recommends that to deter future violence, EUFOR's strength, posture, mobility, and deployments need to be reinforced. Additional troops need to be brought in from EU and non-EU members, while EU/NATO member Pacific Island Countries not presently participating in EUFOR should also make contributions. The restoration of credible deterrence would not only prevent a return to violence, but also may encourage political progress by undermining the capabilities of entrenched elites in manipulating fear. This would create space for citizens and politicians who want the country to function consensually.

The problems with Bosnia-Herzegovina are both practical and psychological. Given that the EU is perceived as weak, divided, bureaucratized, and slow to respond, there are fears that, if the OHR is replaced completely by the EU delegation, the American presence would largely disappear. As a consequence, any effective response to internal violence would be seriously debilitated. This perception itself can encourage political manipulators to test international reaction by stirring division and provoking conflict.

The Sandzak Factor.

Although the Sandzak region, divided between Serbia and Montenegro, did not possess a distinct administrative status or autonomy during the Yugoslav era, the local Bosniak Muslim population has a strong sense of regional identity.[88] According to the 2002 Serbian census, the Slavic Muslim or Bosniak population in Serbia's Sandzak (Raska in Serbian) consists of six municipalities and forms approximately 57 percent of the population of 235,000, or almost 133,000 people.[89] Serbs total fewer than 38 percent of the population, or 90,000 inhabitants. The total number of Slavic Muslims throughout Serbia stands at about 3.5 percent, or 250,000 out of 7.3 million citizens.

For much of the Muslim population in Serbia's Sandzak, Bosnia-Herzegovina is viewed as an ethnic and religious motherland, and such sentiments promote a cross-border Bosniak identity. In an unrecognized referendum in a meeting held during October 25-27, 1991, Sandzak Muslims voted overwhelmingly for territorial autonomy and the right to join Bosnia-Herzegovina. In September 2003, the National Council of the Bosniak Community in Serbia and Montenegro was founded in Novi Pazar, the largest city in the region, adopted a flag and coat of arms, and pushed for the introduction of the Bosniak language in local schools. Serbian officials depicted such moves as separatist provocations.

The establishment of a Bosniak Academy of Arts and Sciences (BANU) by the heads of Muslim communities in Bosnia and Serbia, with dual headquarters in Sarajevo and Novi Pazar, will provide intellectual impetus for a burgeoning national and religious identity and potential unity among Slavic Muslim populations throughout the former Yugoslavia. The creation of BANU has been criticized in Belgrade as a mechanism for increasing ethnic tensions, enhancing the role of Islam in politics, and promoting Sandzak secession.[90] BANU's formation encouraged Croatian activists to establish a Bosnian Croat Academy of Arts and Sciences in Mostar in western Herzegovina. Bosnian Serbs founded their own Academy of Arts and Sciences of the Republika Srpska in Banja Luka in 2006. Some observers believe that these institutions will accelerate Bosnia's division, as intellectuals in all three ethnic groups will operate separately.

A towering figure in Sandzak Muslim politics is Mufti Muamer Zukorlic, head of the Islamic Community of Bosnia-Herzegovina in Serbia. Over the past few years, he has gained popularity by organizing a movement to increase Muslim leverage vis-à-vis Belgrade while campaigning for the region's autonomy.[91] Zukorlic has accused the Serbian government of a slow genocide of the Bosniak population amidst allegedly widespread official discrimination.

To undercut Zukorlic's prominence, the Serbian government engages in a divide and rule policy by sponsoring a rival organization, the Islamic Community in Serbia led by pro-Belgrade loyalist mufti, Adem Zilkic. Zilkic views Hamdija Jusufspahic, the mufti of Belgrade, as the spiritual leader of Serbia's

Muslims. In February 2010, Jusufspahic declared himself the Reis ul-Ulama, or spiritual leader of Muslims in Serbia. By contrast, Zukorlic considers Mustafa Ceric, the mufti of Sarajevo, as the Reis ul-Ulama, and invited him to visit Novi Pazar. Ceric has also visited Prishtina to demonstrate support for Kosova's independence, even though the Bosnian government has not recognized Kosova's statehood.

The Zukorlic and Zilkic factions are at loggerheads, with both claiming to be the legitimate representatives of Sandzak Muslims. Clashes have occurred between supporters of the two leaders in various parts of the region. A similar division exists at the political level between Mayor of Novi Pazar and leader of the List for Sandzak (LZS) Sulejman Ugljanin and Social Democratic Party leader Rasim Ljajic. Ugljanin has also vehemently opposed Zukorlic, viewing him as a political rival. Violent incidents between supporters of the two parties have taken place and precluded Sandzak Bosniak unity by distracting attention from broader national questions.[92]

On October 16, 2011, President Tadic signed an agreement with Bakir Izetbegovic, the Bosniak representative in the Bosnian presidency, according to which Bosniak Muslims living in the Sandzak would be united with Serbian Muslims residing elsewhere in Serbia.[93] The latter group has been traditionally governed from Belgrade in a separate religious administration. The Tadic-Izetbegovic agreement was arranged by Mustafa Ceric, Bosnia's Reis ul-Ulama, and Mehmet Gormez, head of Turkey's state religious affairs directorate. Following the accord, several Bosniak leaders attacked Ceric for betraying their Sandzak brethren to Belgrade. They also criticized Ankara for meddling in Bosniak Muslim affairs and favoring Serbian interests in the region.

Among the demands in Novi Pazar are autonomy in current state configurations or outright separation from Serbia if the RS were to split from Bosnia. There is no strong identification of Bosniaks with the Serbian state but some nostalgia for the defunct Yugoslavia. While Sandzak Bosniaks in Serbia generally supported preserving a single state with Montenegro so that the Islamic population would not be divided, the majority of Montenegro's Sandzak Muslims backed an independent Montenegro so as to remove pressure and discrimination by Belgrade. They calculated that minority rights in the new Montenegrin state would be more far-reaching. The Sandzak was formally partitioned in June 2006 when Montenegro and Serbia gained separate statehoods, thus undercutting any Bosniak Muslim plans to unite the area into one administrative region within a single state.

The Sandzak is one of the poorest areas of Serbia, with high unemployment fueling emigration. More than 50 percent of the economically active population is registered as unemployed—a factor that fosters social and political discontent. Serbian security services allege that Islamic fundamentalism is growing in the region with the establishment of militant organizations run by Wahhabis or Salafis that are the main threat to Serbia's security and that of neighboring states.[94]

Critics of Belgrade believe that Serbian officials deliberately try to divide the Muslim community and to radicalize certain activists to justify crackdowns and anti-Bosniak propaganda. The main Muslim leaders, including Zukorlic, have also criticized the Wahhabi minority for trying to impose its own rituals and beliefs on the Muslim population. For instance, after several incidents in Novi Pazar mosques, Zukorlic banned Wahhabis from praying there.

A few young militants have formed a jihadist group in the Sandzak called Kelimetul-Haqq (Words of Truth).[95] However, Muslim leaders have challenged portrayals of Sandzak radicalization through Wahhabist influence as scaremongering that could further damage inter-ethnic relations. They point out that manifestations of religiosity do not equal Islamic militancy, but such equations by politicians and the official media undermine inter-religious relations and the feeling of safety among Bosniak residents. This was visible in the torching of mosques and attacks on Islamic community centers in Nis, Belgrade, and Novi Sad in March 2004 following attacks on the Serbian minority in Kosova.

While the vast majority of Muslims are moderates, conservative Wahhabism has gained some resonance among a segment of alienated and impoverished youths. In March 2007, the police found "a training camp for terrorists" in Zabren village, some 30 kilometers from Novi Pazar.[96] They arrested a group of young Wahhabi men, who were accused of illegal possession of arms, planning acts of terrorism, and preparations to assassinate Zukorlic. On October 29, 2011, Serbian police arrested 17 people on suspicion of links to the Islamic extremist who opened fire on the U.S. embassy in Sarajevo.[97] Among those arrested, 12 were from three towns in Sandzak with large Muslim communities. Sandzak Muslim leaders have protested against Belgrade's attempt to depict the region as a stronghold of Islamic extremists. However, while not all Wahhabis are political radicals, their stigmatization and criminalization may serve to radicalize them.

The LZS has continued to campaign for the territorial autonomy of Serbia's Sandzak.[98] Serbian laws, including the 2006 Constitution, grant wide-ranging local rights to national minorities, including the right to preserve their language, culture, and identity; benefit from representation in state institutions; and establish minority councils to exercise self-government in specific spheres. However, the implementation of the law has not been consistent in all parts of the country.

Grievances among the Sandzak Muslims center on such questions as under-representation in the public sector, including the police and judiciary, economic stagnation, high unemployment, obstacles to cultural autonomy, and the lack of prosecutions against those responsible for war crimes during the 1992-95 conflict. Such grievances can generate nationalist and religious militancy. Moreover, any moves toward Bosnia's partition will energize proposals for autonomy and separation in the region as a growing number of Sandzak Muslims identify with the Bosniaks and will view the separation of RS as a potential precedent that they can emulate in Serbia.

In the Montenegrin Sandzak, consisting of five municipalities and 164,000 inhabitants, the population is made up of 40 percent Serbs, 27 percent Bosniaks, and 17 percent Montenegrins. In the Montenegrin national census of 2011, 15.97 percent of citizens declared themselves as "from Islam" and 3.14 percent as "Muslim."[99] Bosniak leaders have criticized the range of options available in declaring one's nationality in the Montenegrin census. It serves to dilute and divide the percentage of the population that can be identified as Bosniaks and can undermine their group rights and political representation. Perceptions of discrimination, economic neglect, and under-representation in state institutions can also contribute to national agitation and radicalization among Bosniaks in Montenegro.

IV. REGIONAL ALBANIANISM

The Albanian question remains one of the major unresolved state building puzzles in the Western Balkans and continues to preoccupy international agencies and existing states. Approximately six million Albanians live on the Balkan Peninsula, excluding an estimated two million Albanian descendants in Turkey. They constitute clear majorities in Albania and Kosova.[100] Albania has just over three million; Kosova, 1.8 million; Macedonia, roughly 500,000; Serbia, 50,000; Montenegro, 30,000; and Greece, approximately 750,000. Unlike its Slavic neighbors, the Albanian population is growing and continues to have relatively high birth rates.

The break-up of Yugoslavia was propelled by several unresolved national questions and conflictive national elite ambitions over state structures and the territorial parameters of nations that made up Yugoslavia. Nationalist pan-Albanian movements in the region are small and have benefited from little public support.[101] Albania itself has steered clear of supporting pan-Albanianism, and no major party has such proposals in its platform. Nonetheless, the question could capture the public imagination if a confluence of factors were to crystalize, including prolonged economic distress, alienation from mainstream political parties, growing nationalist appeals, Kosova's destabilization, persistent clashes between Albanians and Slavs in Serbia and Macedonia, stalled prospects for EU membership, and U.S. disengagement from the Western Balkans.

Albania's Turmoil.

Albania has developed a bifurcated two-party system, Democratic and Socialist, despite numerous attempts over the past 2 decades to break the deadlock. Political life is personalized and has been directed by strong leaders where top-down management places limits on intraparty political competition and the input of citizens in decisionmaking. Attempts to form durable third parties have proven difficult. Although some have persisted through several election cycles and enter into government coalitions, the two main parties control over three-quarters of parliamentary seats.

Albania's political disputes are not based on ideologies or programs, as the two dominant parties largely espouse the same goals. Instead, political divisions are grounded within group loyalties among two mutually exclusive political camps. Political clientelism has developed over the past 20 years similar to other Balkan countries and involves an extensive patronage network, a spoils system of official appointments, favoritism shown to supporters of the governing party, and various levels of state-party corruption. Clientelism undermines political competition based on program and merit. It also ensures that political office is lucrative, and losing office is financially damaging and strongly resisted.

Albania has a zero-sum political culture evident during elections. Each ballot is supposed to create clear winners and losers; when the result is close, as witnessed in Tirana's mayoral elections on May 8, 2011, there is little willingness to engage in dialogue and compromise. Instead, the danger persists that disputes will escalate into open conflict. The Tirana elections, with a controversial vote count of misplaced ballots, further accentuated the polarization between the two major parties.

Political contests are not always conducted through elections. Albania has witnessed regular parliamentary boycotts, persistent public protests against election results, and instances of vandalism and violence intended to provoke a government overreaction. According to a European Parliament (EP) report, the violent incidents on January 21, 2011, which led to the death of four demonstrators, exacerbated the climate of mistrust between the two parties and toward state institutions.[102]

As a result of these factors, political confrontations risk escalation into long-term parliamentary boycotts and provocative protest actions. Meanwhile, necessary reforms to meet EU accession criteria are delayed, and the passage of legislation is often blocked. Long-term paralysis will increase social frustration, raise the risk of economic decline, and further erode Albania's qualifications for EU entry. If Albania were to experience prolonged political conflict and social unrest, this could also reinvigorate dormant nationalism as politicians endeavor to gain popularity.

Albania has made limited progress in fulfilling the political criteria for EU membership, including effective parliamentary work, judicial reform, anti-corruption campaigns, a professional civil service, guaranteed property rights, and improved living conditions for the Roma community. The political stalemate has hampered parliamentary work and prevented the establishment of a consensus enabling the implementation of relevant reforms. Opposition boycotts have obstructed the adoption of laws requiring a three-fifths majority.

A potentially negative scenario may unfold through a confluence of negative factors and is more likely to embroil an unstable Albania with limited EU prospects than a politically stable Albania en route to the Union. Such destabilizing elements could include growing social unrest in Kosova as a consequence of international isolation and economic distress that encourage populist and nationalist elements to mushroom; the division of Kosova through unilateral partition supported by Belgrade; a de facto fracturing of Bosnia-Herzegovina that encourages other regional secessions; and political conflicts in Macedonia that assume ethnic dimensions and which would be difficult for Tirana to ignore.

Long-delayed EU accession prospects, combined with economic difficulties, will increase disillusionment with the Union and undermine its effectiveness. Such scenarios could weaken reformist leaders and bring more radical elements to the forefront. They will benefit from economic stagnation and social upheaval and may declare ethno-nationalism and state enlargement as solutions to mounting domestic challenges.

Additional pressures on Albania have been generated by the economic crisis in Greece, where hundreds of thousands of Albanians live and work, providing vital remittances to their families in Albania. In 2007, migrants sent home an estimated $1.3 billion, or approximately 9 percent of Albania's GDP, but by 2010 the total shrank to $690 million and has decreased since then.[103] Many Albanians in Greece have transferred their savings to Albanian banks, fearful that Greece may be forced out of the Eurozone. A growing number of returning migrants will place pressure on the availability of housing and social services. Returnees may not be easily absorbed in the Albanian economy, where unemployment stood at over 13 percent in 2011. Albanian officials estimate that about 250,000 illegal migrants have already returned from Greece in the past 2 years, together with about 15 percent of the legal migrant community, totaling almost 500,000 people.[104]

Nationalist Voices.

The idea of a Greater Albania, or an Ethnic Albania, has been promulgated by some intellectuals but with little political traction or popular appeal. Politically, no Albanian leader in Kosova or Macedonia has been willing to surrender authority to a center in Tirana and become a regional administrator. Albania has not been a magnet of attraction, either economically or politically, for those Albanians who emerged from a much wealthier and open Yugoslavia. Additionally, the international environment was not conducive to Albanian expansionism, especially as American and European restraints on Tirana guaranteed that Albania's political leaders did not play the irredentist card even at the height of the war over Kosova in 1999.

However, the public mood is shifting in Albania, and new actors are appearing on the political stage. In mid-2011, an organization with a quasi-nationalist platform was formed and named the Red and Black Alliance (RBA).[105] By early 2012, the RBA boasted more Facebook members than either the Socialists or Democrats, indicating its increasing appeal among young voters. The organization reportedly also opened branches in western capitals among the Albanian diaspora and claimed to be operating in neighboring Macedonia.

Kreshnik Spahiu, the former Deputy Chairman of the High Judicial Council, leads the RBA. According to his statements, the RBA does not aim to create a "Greater Albania" but merely to foster the rights of Albanians living throughout the region. Its leaders insist that it has no irredentist agenda despite the fact that the Democrat-led Albanian government has depicted RBA as a potential threat to regional security.

The RBA has been building a momentum that other small nationalist parties lack, partly by challenging the two-party establishment, calling for term limits for politicians and parliamentarians, and denouncing pervasive official corruption. Some observers estimate its potential support base at between 25 percent and 40 percent of the Albanian electorate. RBA is tapping into profound public frustration with the political elites and the impunity of official corruption. There is a growing sense of youth alienation from the political class, as the major parties are perceived as serving special interests. Some analysts believe that the RBA could become a king maker in future government coalitions following the next parliamentary elections, scheduled to be held by June 2013.

Both the RBA and the MSD in Kosova deny that they have close ties. Indeed, their leaders appear to be in competition for public support throughout the region and dismiss their rivals either as opportunists, anarchists, Marxists, or folklorists. RBA stresses its pro-American and pro-European credentials and its focus on constitutional action. Nonetheless, Albania may experience growing disillusionment with international representatives, especially if progress toward EU membership is indefinitely delayed. This can be coupled with a sense of frustration with the foreign diplomatic presence that appears incapable of dealing with perceived electoral fraud and widespread official corruption.

Religion has not determined Albanian politics, as the nation does not identify itself with a specific denomination. Although the majority of Albanians are nominally Muslim, there is little sense of a common Muslim religious identity with non-Albanian Islamic populations. Albanian nationalism has always been secular, as Albanian leaders sought to avoid divisions between members of the Muslim, Catholic, and Orthodox faiths in the program of nation building.[106] Although Islamic organizations have raised their profile in Albanian societies, especially in Macedonia and Kosova, conditions are not propitious for any significant growth in radical Islam. In fact, no political party of any significance has rallied around Islamic doctrine or symbols.

In Albania, religion has not figured as a source of conflict, but recent developments in the Islamic community need to be monitored. Muslim leaders have been embroiled in a dispute since early 2012 following the dismissal of Lulzim Plloci, imam of the Madrasa Mosque in Tirana.[107] The move was perceived as part of an attack by the leadership of the Muslim Community in Albania (MCA), which favors a Turkish brand of Islam rather than using clerics educated in Arab countries. The League of Albanian Imams, a splinter group from the MCA, condemned Plloci's dismissal, as it favors an Arab brand of Islam. The split between the two groups began in the wake of September 11, 2001 (9/11), when the government examined the background of foreign Islamic charities operating in Albania. Turkish Islam is regarded as culturally more in tune with Albania's Muslims and less vulnerable to radicalism, but the MCA ruling has generated conflict over Islamic leadership.

Kosova's Aspirations.

Social frustration is mounting in Kosova, as evident in the electoral emergence of an opposition group with a program that combines anti-corruption, full national sovereignty, and a pan-Albanianism that could resound more broadly among the frustrated and underemployed citizenry. Social unrest can be compounded by Prishtina's relative international isolation, its long-term exclusion from the EU, and Belgrade's push toward unilateral partition of the new state. Although political dissatisfaction does not currently revolve around ethno-nationalism, it can assume such forms if Kosova becomes viewed as a "frozen state" blocked from entering the major international institutions.

Key reforms have been delayed in Prishtina, especially in public administration and the judiciary and in tackling organized crime and corruption. The EU has called for a more proactive approach by law enforcement agencies and judicial authorities, while its monitoring and mentoring mission, EULEX, has also come under criticism for its shortcomings in promoting the rule of law. In the northern municipalities where Prishtina has no control, access to justice is not fully guaranteed, notably in northern Mitrovica where the district court functions with limited capacity.

After 4 years of outside supervision, Kosova's parliament adopted a resolution on January 31, 2012, calling for the international community to close the International Civilian Office (ICO).[108] The International Civilian Representative for Kosovo, supported by the ICO, is also the EU Special Representative (EUSR) for Kosovo appointed by the Council of the EU and the final authority in interpreting the Ahtisaari Plan. The Kosova authorities pledged to meet all obligations under the Ahtisaari Plan that imposed supervision over the country's institutions. On September 10, 2012, the International Steering Group (ISG), which oversees Kosova's independence, endorsed closing the ICO mission while seeking firmer guarantees from Prishtina in such areas as decentralization and minority rights. This decision will enable Kosova to demonstrate whether the state is increasingly functional or inherently unstable.

EU monitoring through the EULEX rule of law mission, intended to help Kosova develop European standards, expired at the end of 2012, although it was extended for another 2 years under the name "EULEX Kosovo." EULEX has acquired a mixed image. Supporters in Brussels argue that it has helped reduce crime and succeeded in training local police and customs officials.[109] Critics contend that although the EU has investigated several corruption cases, it has not enabled the conviction of any high-level offenders, and the local judiciary has not improved its operations. Paradoxically, a more forceful EULEX role against organized crime and official corruption could undermine political stability as it may implicate some members of the government.

The EP has periodically criticized EULEX and urged the mission to increase its efforts against organized crime and terminate road blockades by Serbian activists in northern Kosova. The government in Prishtina wants the EULEX role limited to dealing with war crimes and international crime, while it takes over full responsibility for justice and internal policing. The biggest test for Prishtina will be to integrate its four northern municipalities where Serbs are campaigning for autonomy or secession.[110] In synchrony with EULEX, the Kosova government has drawn up plans to fully incorporate the northern region, including the creation of a Temporary Administrative Office for North Mitrovica, the formation of transitional local governments in the four northern municipalities, and the closure of separate Serbian institutions.

The ISG is committed to Kosova's territorial integrity and has urged the Serbian authorities to withdraw their security units from the country. Although Belgrade denies having any official presence in the region, Prishtina claims that Serbian Interior Ministry personnel are present and have links with local militants and organized crime networks.[111] ISG asserts that if Belgrade took a more constructive role, it would make it easier for the rule of law to function in Kosova and would accelerate the transfer of power from international actors to the country's authorities.

In the northern municipalities, neither the Kosova police nor the Kosova Security Force can fully perform their duties given the presence of militant opposition and a lack of cooperation from Belgrade. Indeed, the evacuation of EULEX may provide the Serbian authorities with a valuable opportunity to demonstrate that Kosova is not a viable state by undermining its territorial integrity. Serbian majority municipal assemblies in the north staged a referendum on February 14-15, 2012, to decide whether Serbs wanted to be part of Kosova. Turnout was 75.28 percent, with 99.74 percent reportedly rejecting Prishtina's authority.[112]

One additional complication is the ongoing UN mandate over Kosova, which remains valid in the area of rule of law. In November 2008, the UN Security Council allowed the Secretary General to transfer UN responsibility for the rule of law to EULEX. If UN resolution 1244 is not replaced, responsibility for the rule of law in Kosova will reportedly revert to the UN, should EULEX evacuate. This could mean the return of UN police and judges to the north, a scenario that will be unacceptable to Prishtina and could culminate in conflicts with the international presence.

Pan-Albanian Opportunities.

Support for a Greater or Ethnic Albania can escalate as Kosovars voice frustration with their country's limited progress toward membership in international institutions, become dissatisfied with international mediation, distrust state institutions and political elites, and continue to suffer from economic underdevelopment. The economic situation remains precarious, and the government needs to take urgent steps to improve the budgetary situation in close coordination with the IMF.[113]

Kosova's most severe test may come from within the Albanian population as the youthful MSD expands its popularity and exacerbates its disputes with the government. Led by Albin Kurti, a former Albanian student leader in Kosova during the Milosevic years, MSD emerged on the political scene during the December 12, 2010, general elections. The movement captured almost 13 percent of the national vote in its first ballot. It finished third in the elections and gained 14 out of 100 parliamentary seats.

MSD has been outspoken in its criticisms of the Kosova administration and of foreign representatives in Prishtina. Its program revolves around two major planks. First, it contends that Kosova is governed by a corrupt and incompetent political elite, which is incapable of developing a modern and economically prosperous state or gaining full self-determination for the country. Second, MSD charges that international representatives in Prishtina simply maintain the political status quo, fail to uphold Kosova's territorial integrity, and are insufficiently active in gaining Kosova international recognition and membership in multinational organizations. The EULEX mission is berated for its inability to help construct a modern state. Such criticisms have led the MSD to oppose what it believes is counterproductive foreign interference.

In addition, Kurti has accused the coalition government of Prime Minister Hashim Thaci of engaging in talks with Belgrade that bring no benefits to Kosova but may actually delay the country's progress. MSD activists have staged several protest actions, including the blockade of border crossings with Serbia. MSD spokesmen asserted that the blockades were in defense of the domestic economy, as Serbia floods Kosova with goods while blocking Prishtina's exports. It was also intended to highlight the frontier as an interstate border and not just an "administrative line" as claimed by Belgrade.

MSD has the potential of both rejuvenating and radicalizing Kosovar politics. Its combination of parliamentary pressure and street politics—together with its growing popularity, outspokenness, and confrontational stance—has unnerved international actors. Kurti has come under criticism from several EU representatives. The EP rapporteur for the Western Balkans Jelko Kacin charged that clashes between Kosovar police and MSD protesters during the summer of 2012 damaged international attempts to improve relations between Belgrade and Prishtina and undermined the EU-sponsored talks with Serbia.

Kacin accused the MSD of acting the same way as Serbian minority leaders in northern Kosova who have blocked border posts to protest against the presence of

Kosovar police and customs officials. Kacin rejected Kurti's explanation that the protest was a reciprocal measure against Serbia's failure to implement trade and other agreements with Kosova. In an evident attempt to disguise the EU's own shortcomings, Kacin claimed that the border clashes retarded international efforts to persuade Belgrade to end its obstruction of Kosova's participation in regional initiatives.

The MSD was energized by the February 24, 2012, agreement between Belgrade and Prishtina on representation in regional organizations.[114] According to the deal, Kosova would be represented simply under the name "Kosovo" and not the Republic of Kosova, and with an added footnote that provoked widespread outrage. The wording of the footnote challenged Kosova's independence and can be used to undermine its statehood. It read: "This label [Kosovo] does not prejudge the status of Kosovo and is in accordance with Resolution 1244 and the opinion of the International Court of Justice on Kosovo's declaration of independence." Logically, if the status of Kosova is not "prejudged," then it is unsettled: the position the Serbian government has maintained since the country declared independence in February 2008. The footnote phrasing and its interpretation contradict Kosova's constitution and statehood, which the United States and most EU countries have pledged to uphold.

Subsequent arguments erupted on whether the Kosova footnote should be mentioned only in agreements and official documents and not on nameplates at meetings. Another complication concerned the spelling of the country's name—Kosova (the Albanian version) or Kosovo (the Serbian version). The agreement reached in Brussels did not specify how name plates should be written and opened up the terrain to conflictive interpretations.[115]

Prime Minister Thaci was under intense pressure from Washington and Brussels to sign the footnote agreement. He tried to put a brave face on the deal by claiming that the designation was temporary, but his domestic credibility suffered, as there was no indication that the wording would be altered in the foreseeable future. Thaci claimed that through the EU-brokered deal, Belgrade had effectively recognized Kosova's statehood because Prishtina could participate in regional fora in which Serbia is a member.

While the government in Prishtina claims the footnote agreement will facilitate relations with EU members that have not recognized Kosova, opposition leaders assert that it will damage Kosova's international status, as many countries will remain neutral on the question of status. The MSD described the concessions made to Serbia as a serious setback, threatening Kosova's sovereignty and international status. It called for street protests and challenged the personal position of the Prime Minister, who had failed to discuss the footnote accord in parliament.

In addition to reviving Serbian nationalist claims, the footnote agreement partially lifted the lid on a simmering pot of Albanian nationalism. Until now, Kosovars have been grateful to international players for their national liberation and independence. But a more self-confident generation is emerging that no longer feels beholden to foreign powers. They are neither anti-American nor anti-European but increasingly focused on Albanian and Kosovar interests and less willing to compromise on basic principles.[116]

Outgoing U.S. ambassador to Kosova Christopher Dell condemned those objecting to the outcome of the talks in Brussels as "anti-American and anti-European." Such simplified labeling no longer scares young Albanians but may create even more resentment against foreign interference in domestic politics and perceptions of Kosovar submissiveness. Moreover, any government crackdown on MSD, coupled with perceptions of official backtracking on statehood, could raise the movement's support base. In March 2012, Kurti claimed that MSD could gain over 21 percent of the vote in upcoming elections.[117]

Neither Washington nor Brussels seem prepared for a rising wave of Albanian self-assertion that will not simply manifest in street protests and blockades but may be increasingly reflected in support for nationalist parties that are less willing to follow Western recommendations. International decisions that are seen as sacrificing Kosovar aspirations for the sake of neighboring capitals will be resented and may spark even more expansive nationalist demands.

The MSD possesses a pan-Albanian agenda that favors unification between Kosova and Albania. Although it is not actively engaged in such a process, the mere fact that its leaders openly discuss unification as a viable future option has unnerved political parties and international actors who have avoided or dismissed the question for over 20 years. MSD has proposed a referendum on Kosova's unification with Albania, an idea that is gaining traction among the younger generation in both countries. It considers such a merger as creating a stronger and more viable state. In pursuit of this project, observers claim that MSD is developing links with Albanian nationalist and unification groups in neighboring countries.[118]

As a younger generation of political activists comes to the forefront, cross-border Albanian politics will become increasingly interconnected. For instance, Kurti is a popular figure in Tirana and is considered to be one of the most articulate Albanian leaders. His appeals to the wider Albanian nation could stir the specter of pan-Albanianism that all major political parties have eschewed since the collapse of communism and Yugoslavism.

In denigrating Kosova's aspirations toward independence, Belgrade has manipulated Islamic fundamentalist and terrorist stereotypes that carry resonance in the West. The term "Wahhabi" is widely used in the region not as an accurate depiction of a specific religious community but as a label to discredit one's political opponents. Although some Wahhabist groups have been active in parts of Kosova through charity work and the restoration of mosques, their ideology has limited public resonance or political impact.[119]

Nonetheless, national radicalization in Kosova could also encourage a turn to religious conservatism among some segments of the population.[120] Observers cite efforts by the conservative Justice Party to amend the constitution, which declares Kosova a secular state, to allow hijab in public schools, and to construct a large mosque in Prishtina that would absorb the growing numbers of worshipers. Some Kosovars attribute rising piety among poorer sectors of society to Muslim charities. Several Muslim NGOs operate in Kosova rebuilding mosques destroyed during the war and offering financial help to orphans. Muslim NGOs are also engaged in health and educational projects, with critics accusing them of using their influence for ideological inroads among poor Kosovars.

More troubling for regional security has been the re-emergence of secular guerrilla groups, such as the Albanian National Army (ANL), which has been designated as an illegal and terrorist organization by the UN and the United States. It is believed to have close links with the Front for Albanian National Unification, a group that advocates the unification of Albanian majority territories in Albania, Kosova, Macedonia, Montenegro, Serbia, and Greece.[121] In March 2012, the ANL claimed that it had reactivated its structures to protect Kosova from a Serbian invasion.

Another clandestine group, the Army for the Liberation of Occupied Territories (ALOT), announced its existence in April 2002 claiming that Serbs were intent on truncating Kosova.[122] ALOT claimed responsibility for spraying a Serb vehicle with bullets in the village of Cabra north of Mitrovica. Serbia's Minister for Kosova, Goran Bogdanovic, urged KFOR to disarm Albanian militants, claiming that their extremism could lead to escalating violence.[123] Local Serbs also felt intimidated after flyers were distributed in several villages calling for their expulsion. The flyers were reportedly signed by the ANL.[124]

UN Secretary General Ban Ki-moon warned in his May 2012 report on Kosova that tensions and confrontations constitute a serious risk to stability. Ban noted that the number of crimes against ethnic minorities increased between February and May 2012, relative to the same period the previous year. Threats, thefts, arson, vandalism, and attacks on the facilities of the Serbian Orthodox Church were on the rise.[125] Equally troubling, the Serbian ultra-nationalist organization Obraz reportedly pinned threatening messages on the doors of Albanians living in northern municipalities. Local Albanians subsequently demanded an increased security presence in their neighborhoods. In response, NATO's peacekeeping mission to Kosova relocated some troops to mixed ethnic areas. Police sources in Mitrovica reported that the Obraz leaflets raised feelings of insecurity within the Albanian community.[126]

Presevo Valley Conflicts.

According to the 2002 Serbian census, Albanians form a majority in two municipalities of southern Serbia, Presevo (89.09 percent) and Bujanovac (54.69 percent), and a sizeable minority in the municipality of Medvedja (26.17 percent). All three municipalities border Kosova. Unless the emigration of Albanians accelerates, longer-term demographic trends do not favor the Serbs.

The region experienced a 17-month insurgency in 2000-01 following the liberation of Kosova by NATO troops during Milosevic's campaign of expulsion and mass murder, and with several atrocities by Serb militias reported in the Presevo region. The insurgency ended in May 2001 with the involvement of NATO, the United States, and the Organization for Security and Co-operation in Europe (OSCE) and the signing of the Konculji Agreement between the Liberation Army of Presevo, Medvedja, and Bujanovac (LAPMB) and the Serbian administration. About 100 people were killed during the fighting, while over 12,500 Albanians fled to Kosova.

Under the Covic Plan, in return for disarming and demobilizing, the guerrillas were to be amnestied, refugees would be allowed to return, a multiethnic police force would be formed, and Albanians would be integrated into state institutions after decades of exclusion and discrimination.[127] Belgrade established a

Coordination Body (CB) for the three municipalities to defuse ethnic tensions, allow refugees to return home, and support local economic development. However, smaller-scale incidents continued, and tensions have persisted. Former LAPMB fighters maintain close contacts with ex-KLA guerrillas in Kosova. Some Serbian officials continue to view the Albanian minority as a separatist element that endangers the country's north-south transportation corridor, including the probable route of future south-north energy pipelines.

In January 2006, Albanian leaders in Presevo, Bujanovac, and Medvedja called for far-reaching decentralization and autonomy in the Presevo valley, reminiscent of Serbian demands in northern Kosova. They also adopted a common platform calling for the unification of the Presevo valley (or Eastern Kosova) with Kosova in the event of changes to Kosova's northern borders.

In September 2007, representatives of the valley's five largest Albanian parties issued a declaration in support of Kosova's statehood and called for a greater international presence in the Presevo region. The danger persists that anti-Serb violence in Kosova could precipitate anti-Albanian violence in Presevo, and vice versa. If the separatist agenda for northern Kosova is pursued by Belgrade and gains some international backing, two options would emerge: exchanging territory between northern Kosova and southern Serbia, or exchanging minority populations between the two subregions.

Any official statements by Belgrade regarding Kosova's partition or the autonomy of northern Kosova creates uncertainty and tension in the Presevo region. In the event the partition of Kosova continues to be firmly rejected by international players while the northern municipalities evade Prishtina's jurisdiction, support may grow for the transfer of Serbs from northern Kosova to the Presevo valley. This would alter the ethnic balance in the valley in favor of Serbs, provoke conflicts with resident Albanians, and stimulate calls for the reciprocal transfer or expulsion of Albanians from the Presevo valley.

Belgrade has registered some progress in Presevo by allowing for the formation of multiethnic local governments, joint Serbian-Albanian police patrols, and improvements in the Albanian-language media. In January 2007, Albanians elected a representative to the Serbian parliament. However, limited headway has been made in educational reform and the integration of Albanians in state institutions, including the judiciary, and tensions persist between local residents and police units. Albanian leaders also criticize Belgrade's CB, which has failed to deliver on various promises of reform and is seen as an arm of the Serbian government rather than a consultative organ.

Dissatisfaction with Serbian rule is perpetuated by harsh economic conditions, including high unemployment, poor infrastructure, and general impoverishment, especially in rural areas. Trade between the region and Kosova and Macedonia has also been restricted because of Belgrade's security fears. Serbian forces along the border conduct stringent checks on traffic and disallow Kosovar citizens from crossing into the region. Concerns about security have also discouraged internal and external investment in the Presevo valley municipalities.

Albanians in southern Serbia massively boycotted the October 2011 census, thus undermining its accuracy and legitimacy. Splits are also visible among Albanian parties, with some adopting more radical positions and being less willing to seek compromises with Belgrade. In the event of increased tensions, the more militant factions are likely to raise their popularity and compete over pan-nationalist agendas such as unification with Kosova.

Several additional developments have increased tensions in southern Serbia. The announcement in October 2010 that a Serbian-Russian center for "emergency coordination" would be built in Nis near the Presevo region may indicate official anxiety about Albanian unrest, while demonstrating that any moves toward autonomy will be prohibited. Russian officials inaugurated this "regional humanitarian center" on October 17, 2011, by claiming that it would contribute to a more efficient response to emergencies not only in the Balkans but also throughout Europe. They denied suggestions that Moscow was establishing a military base.[128] Nonetheless, speculation has persisted that the Nis facilities could be turned to military use.

On the eve of the Serbian elections on May 6, 2012, five Albanians were arrested in the region on suspicion of committing war crimes during the 2001 Presevo rebellion.[129] Conducted in the town of Bujanovac and in the villages of Veliki Trnovac and Breznica, these arrests heightened tensions in the volatile area. A local Albanian politician described the arrests as an act of "state terror" against ethnic Albanians.[130]

Former Albanian paramilitaries issued warnings in May 2012 that they would restart their operations in the region if arrests of Albanians continued.[131] Jonuz Musliu, former political chief of the Liberation Army of Presevo, Medvedja, and Bujanovac, claimed that Belgrade was avoiding a peaceful solution to the conflict. Albanian leaders in the Presevo valley stated that their biggest concern was the warning by Interior

Minister Ivica Dacic about continuing police actions. Bujanovac Mayor Shaip Kamberi claimed that Belgrade's goal was to destabilize the region and intimidate the Albanians to leave en masse.

Macedonian Pressures.

The spillover from the armed conflict in southern Serbia contributed to the outbreak of insurgency in Macedonia in August 2001, as well as to several clashes between Albanian guerrillas and Macedonian government forces since then. The Ohrid Framework Agreement, negotiated under international supervision in August 2001 between Slavic and Albanian Macedonian representatives to provide Albanians with more significant representation in state institutions and broaden minority rights, was largely implemented, but new frictions have arisen in recent years. These include disputes over the allocation of state resources and protests against the fervent nationalism of the current Macedonian administration.

Albanians complain that the percentage of minority civil servants does not correspond with the Ohrid stipulations. In 2011, the share of Albanians reached 17.2 percent, short of the goal of 25 percent representing the estimated percentage of Albanians in the country. The figure has barely increased in recent years.[132] In addition, in March 2012, the Macedonian parliament turned down a request by Albanians to add Albanian Flag Day to the national calendar, thus disappointing the minority community and further estranging them from the state.[133]

Islam has a more visible presence among Albanians in Macedonia than elsewhere in the Balkans because of the growing role of the Macedonian Orthodox

Church among Slavic Macedonians in their assertion of national identity. The urban renewal movement in Skopje has been closely tied to the Slavic Macedonian national renaissance and the role of the Macedonian Orthodox Church. This has alienated many Albanians, who increasingly view the official Islamic Community as a defender of their interests.

Several faith-based organizations and Islamic "missionaries" from Saudi Arabia and the Gulf have been active in Macedonia and Kosova, preaching a conservative brand of Islam. In Macedonia, this has contributed to a struggle within the Islamic population between the moderate mainstream of the Islamic Religious Community and pockets of pious Wahhabis influenced from abroad. However, it has not resulted in the radicalization of the majority of religious believers or in the intrusion of religion into political life. In both Kosova and Macedonia, Albanian leaders do not want their populations radicalized as they seek membership in both NATO and the EU. According to local analysts, the surest way to prevent the growth of extremist religious ideologies is to improve living standards and instill a social safety net, which restricts opportunities for militant proselytizers.[134]

Ethnic clashes have erupted in Macedonia, as witnessed in January 2012 when both Albanian and Slavic Muslim communities expressed outrage over a carnival in which Orthodox Christian men mocked Muslims by dressing as burqa-clad women.[135] The incident at the Vevcani festival prompted demonstrations in several Macedonian towns and expressions of anger against Macedonians during sports events in neighboring Kosova. Incidents of hostility between Albanian and Macedonian supporters have become a regular feature of sports events in Skopje and Tetovo.[136]

Reports about the revival of guerrilla groups also surfaced in Macedonia during 2012. Xhezair "Commandant Hoxha" Shaqiri of the National Liberation Army (NLA), an offshoot of the KLA, asserted that the organization's former commanders were considering remobilizing their troops if the "provocations continue."[137] In addition to internal Macedonian conflicts, the unsolved status of northern Kosova was generating disquiet over territorial partition with potential implications for Macedonia.

Recent opinion polls indicate that two-thirds of the residents of Albanian-majority districts in western Macedonia support the creation of a common Albanian state with Albania and Kosova, and more than half think it will soon materialize.[138] Although no active political mobilization for separation is underway in the country, this could change if relations between Albanians and the ruling Macedonian party were to deteriorate. Polls conducted by the Skopje-based Center for Inter-ethnic Tolerance before tensions escalated in 2012 revealed that 78 percent of respondents believed that inter-ethnic relations were very bad, and 71 percent considered inter-ethnic intolerance to be on the rise.[139]

Montenegrin Dimensions.

According to the Montenegrin census of 2011, Albanians account for 4.9 percent of the country's population of 620,000, or approximately 30,000 inhabitants.[140] They are concentrated close to the Albanian border and form a majority of 70.66 percent in the municipality of Ulcinj and just over 50 percent in the municipality of Tuzi. Although most Montenegrin Albanians are Muslims, there is a substantial minority of Catholics, similar to northern Albania.

The Montenegrin government honors several provisions to protect minority rights. For instance, Albanians have access to the minority-specific broadcasting media, while a set number of parliamentary seats are allocated to Albanians and other minorities. In 2005, a total of 11 members of ethnic minorities were elected to the 75-seat parliament and three minority members appointed to cabinet positions. By 2006, the number of minority members in the 81-seat assembly reached 14, although their number in the cabinet dropped to two. In 2007, their political representation improved, with 16 minority members in the assembly and two in the cabinet.

Despite these positive indicators, Albanian leaders complain that their community continues to suffer from discrimination and neglect.[141] For example, although Podgorica funds Albanian-language education in local primary and secondary schools, together with some university-level courses, Albanians still campaign for equal rights to use their own language and develop their education systems in areas where they predominate. They claim that text books ignore Albanian history and the physical condition of schools is appalling.

Many Albanians voice disappointment that despite supporting Montenegrin independence, their situation has not improved, and their population has dropped through emigration. Economic conditions have stagnated, and the country's privatization process has contributed to social dislocation, economic hardship, and official corruption. All these factors, together with the weakening economy, have exacerbated inter-ethnic tensions, as each community fears it will lose access to scarce resources.

Albanian leaders campaign vigorously for administrative decentralization, especially in Ulcinj municipality. They complain that government measures have stripped the municipalities of their authority over justice, education, health, and local police. Among other grievances are the lack of funding for cultural activities; the absence of national institutions to develop folklore and ethnography; and no national theater, arts gallery, publishing houses, media centers, or national institute devoted to preserving the Albanian language and culture. Additionally, Albanians complain about a lack of access to government jobs. Whereas over 20 percent of Slavic Montenegrins are employed by the state, the total is under 10 percent for Albanians, and they are particularly under-represented in justice and internal security.

Although Montenegrin Albanians have not voiced any secessionist demands, their extensive list of grievances could contribute to breeding dissatisfaction and radicalize the political scene. It will also feed into the pan-Albanian arguments and aspirations of rising political leaders in neighboring Albania and Kosova.

Chameria vs. Epirus.

Small nationalist groups in Albania have periodically raised the question of the Chameria territory in northern Greece; and in 2011, a new party was formed, the Justice, Integration and Unity Party (PJIU), that focuses primarily on Cham demands.[142] It gained two seats in parliament and joined the Democratic Party-led coalition government. An estimated 200,000 Cham Albanians are mostly descendants of those expelled from what became northern Greece after the Balkan wars of 1912-14, following the signing of the Turkish-

Greek Convention at Lausanne in January 1923 and after World War II. Cham movements in Albania want Athens to account for about 4,000 Chams who disappeared as a result of the conflicts, as well as compensation for property seized from approximately 150,000 Chams. About 40,000 Christian Orthodox Cham Albanians, mostly old people, still reside in Greece.[143]

The RBA and PJIU have been vociferous regarding the Cham question and the treatment of Albanian immigrants in Greece. They also issue periodic warnings about Greek separatism in southern Albania. They called for a boycott of the October 2011 census and opposed the law for changing one's nationality, arguing that thousands of Albanians have declared themselves as Greek over the past 20 years simply to reside and work in Greece, which has inflated the size of the Greek minority in Albania.[144] Such a process allegedly encourages Greek irredentism toward southern Albania, which Greek nationalists consider to be Northern Epirus and thereby a part of Greater Greece. Some Greek minority activists, together with nationalists in Greece itself, have called for autonomy for the southern Albanian region. Nationalist fervor is also raised by difficulties in reaching an agreement on the maritime border between the two states.

Ethnic minority representatives, including Greeks, Macedonians, Serbs, Montenegrins, Vlachs, Roma, and Egyptians, announced in December 2011 that they will pursue a lawsuit at the European Court of Human Rights in Strasbourg following Albania's Constitutional Court ruling to erase the category "nationality" in legal and civil registries.[145] The court determined that nationality is not necessary to include on the census forms to enumerate the population. The Albanian court acted on the legal challenge issued by several local judges and by the RBA, one of the main opponents of recording ethnicity and religious identity in the Albanian census.

RBA leaders claim that leaders of the Greek minority may demand special status and a percentage of parliamentary seats and government positions, along the lines of the Ohrid model in Macedonia or the Ahtisaari plan in Kosova. Conversely, minority leaders allege that the Albanian government is openly discriminating by making it unconstitutional for citizens to be members of any nationality other than Albanian. Representative of the Greek minority Unity Party for Human Rights also condemned international representatives in Tirana for ignoring minority demands. Subsequently, several minority leaders signed a joint declaration refusing to recognize the results of the 2011 census.

On July 1, 2012, some 4,000 members of Cham organizations held a protest near the Greek-Albanian border in Thesprotia demanding the abolition of Greek laws that prevent them from accessing their traditional lands and properties in Greece.[146] The leader of the PJIU, Shpetim Idrizi, accused Greece of violating human rights by not allowing Chams to visit their ancestral homes. A Genocide Monument has been constructed by Chams in the southern Albanian town of Sarandë, with plans to construct a larger monument to the victims of genocide.

The Chameria question has been largely dormant for 20 years, as Tirana did not want to antagonize Athens during its campaign to join the EU and NATO. Nonetheless, Tirana has recently taken the issue to the World Court of Justice in the hope of gaining some financial compensation for confiscated Cham property. Cham activists in Albania who have become increasingly outspoken are campaigning for several specific actions, including a resolution in the Albanian parliament obliging all governments in Tirana to speak up on the Cham question, return of property and land seized during the expulsions or appropriate compensation payments, and Greek citizenship for expellees and their descendants so they can either return to their ancestral areas or visit them freely.[147] Cham activists claim they have no irredentist pretensions to northern Greece even though officials in Athens dismiss them as radicals and separatists and reportedly prevent them from travelling to the Chameria region.

Cham activists complain that Albanians and other minorities in Greece are denied group rights in such areas as education, language, or religion, in contrast to the broad array of minority rights granted to the Greek community in southern Albania. The Greek population is officially estimated at under 25,000, mostly residing in 99 villages in southern Albania, with their own political organization, the Unity for Human Rights Party (UPHR) with a seat in parliament. President of the UPHR Vangelis Dule unleashed a storm in March 2012 during an interview with an Albanian TV station by claiming that there was no Chameria region in Greece and that the Cham question was a fabrication.[148] He accused several Albanian deputies of nationalism, and his life was subsequently threatened.

Reports periodically surface about the existence of a guerrilla movement styled as the Chameria Liberation Army, which seeks to create a Greater Albania through insurrection. The rumors appear to have more propagandistic value than actual substance. However, the Cham problem has been taken aboard by nationalist and pan-Albanian groups in Albania as part of their program to defend the rights of Albanians in neighboring states. The addition of Chameria to a prospective Ethnic Albania project could become more enticing for various pro-unification groups.

Greece could be on the verge of a social explosion that could have an impact on its neighbors. Whether Athens defaults on its massive debts or qualifies for new international loans, the country faces unprecedented social turmoil. Ultra-leftist and ultra-rightist parties benefited in the June 17, 2012, parliamentary elections and are supported by over 40 percent of the Greek electorate. The nationalistic and openly chauvinistic Golden Dawn party gained almost 7 percent of the vote and 18 parliamentary seats.

Public frustration in Greece is intensifying, as the government has pledged to make deeper cuts in the minimal wage, pensions, and state sector jobs to reduce the budget deficit and obtain emergency funds from the EU and IMF. The voters' revenge could result in a series of weak governments or propel to power a more radical coalition. In another destabilizing alternative, Greece may become insolvent and leave the Eurozone with a comprehensive decimation of living standards. Unemployment among young people reached 48 percent in mid-2012, and GDP has contracted by 20 percent during the 5-year recession. Meanwhile, the bloated state sector and powerful trade unions continue to block structural reforms necessary for economic recovery.

Greece's social breakdown may become comparable to Albania's in 1997 when the financial pyramid schemes collapsed. But instead of leading to potential state failure, Greek turmoil can presage the emergence of an authoritarian government that will freeze the country's democracy. Under the pretext of restoring order and defending Hellenic dignity, a nationalist regime could pinpoint internal and external enemies to deflect public rage. Minorities can become especially vulnerable to attack and be forced to leave the country. This can provoke conflicts with directly affected neighbors.

The most obvious external targets for Greek nationalism would be Turkey, Macedonia, and Albania, where it could provoke equally nationalistic reactions. Conflicts between Ankara and Athens would affect security in Cyprus, the Aegean, and the Balkans. Athens can also rekindle the northern Epirus (southern Albania) question to divert attention from internal turmoil. On February 20, 2012, during demonstrations in Athens against government austerity measures, a delegation of Greeks from southern Albania laid a wreath at the Tomb of the Unknown Soldier in commemoration of the anniversary of the declaration of the Autonomous Republic of Northern Epirus in 1914.[149] The ceremony demonstrated that the manipulation of historical anniversaries and territorial grievances by political leaders on both sides of the border should not be underestimated.

V. MACEDONIAN IMPASSE

The Macedonian state has stalled in its moves toward membership in both NATO and the EU, and this is both a cause and a consequence of rising national assertiveness. Several dangers lurk ahead for Skopje if the country's progress into NATO and the EU remains indefinitely blocked.[150] The absence of a solution to the ongoing name dispute with Athens will rebound negatively on Macedonia's political stability, undermine economic development, diminish foreign investment, halt progress in necessary structural reforms, intensify manifestations of ethno-nationalism, and potentially lead to escalating inter-ethnic conflicts that will endanger the country's territorial integrity and challenge regional security.

Nationalist Resurgence.

Since it declared independence from Yugoslavia on September 8, 1991, the Republic of Macedonia, according to its constitutional name, or the Former Yugoslav Republic of Yugoslavia (FYROM), according to its designation in major international institutions, has made substantial progress in transforming itself into a contender for both EU and NATO membership. In particular, following the brief Albanian insurgency in the summer of 2001 and with intense Western involvement, significant steps were undertaken by the government to integrate the large Albanian minority into the country's institutions. However, the path to both NATO and the EU has not proceeded smoothly because Macedonia needs to resolve its dispute with Greece over the country's name and other national identifiers, as this has become a primary condition for incorporation in both the Alliance and the Union.

On June 5, 2011, the Internal Macedonian Revolutionary Organization-Democratic Party of Macedonian National Unity (VMRO-DMNE) was re-elected and formed a coalition government with the major Albanian party, the Democratic Union for Integration (DUI). VMRO leaders have demonstrated little willingness to make concessions over the name dispute with Athens, so that Macedonia's progress toward both NATO and EU membership stands at an impasse. Indeed, VMRO's populist credentials in staunchly defending Macedonian national interests are strengthened by its opposition to any compromise with Greece over the country's name, the definition of its people, and the designation of its language.

In addition to the dispute with Greece, Macedonian and Bulgarian authorities have sparred over VMRO's appropriation of Bulgarian Tsars as ethnic Macedonian rulers, even though the historical record does not mention Macedonia as an administrative structure or a distinct national identity until the end of the 19th century. For instance, a statue has been erected in the middle of Skopje to Bulgaria's Tsar Samuel, the ruler of the First Bulgarian Empire (997-1014 AD), to link him with a purportedly longer Macedonian historical identity. Such moves and the "Macedonianization" of Bulgarian history in school textbooks have provoked angry exchanges with Bulgarian government representatives.

To preserve their influence and power, and riding on a nationalist wave, VMRO leaders have expanded the state administration to reward party supporters. The civil service has been transformed into a largely partisan network driven by nepotism, patronage, and clientelism.[151] The party seeks control over various key social sectors, including business, the media, the health system, academia and education, trade unions, NGOs, and professional organizations. State expenditure on administrative expansion and prestige projects in the capital have damaged the national budget and curtailed investment in national infrastructure. These problems are compounded by widespread corruption and mismanagement, stalled prospects for EU and NATO entry, and global economic trends that have squeezed economies such as Macedonia's, which remains dependent on foreign investment and assistance.

Macedonia's opposition parties have regularly protested against the extensive renovation work that has dramatically altered the center of Skopje, erected kitschy statues and other expensive monuments in the capital, and foresees the renaming of hundreds of streets after Macedonian national heroes. The opposition Social Democratic Party accuses the authorities of rewriting history and selecting figures from the past that conform with VMRO's nationalist ideology.[152]

Despite its hardline stance, time appears to be working against Skopje for several reasons. First, EU countries have either lined up behind Greece as a co-member or have remained neutral and will not support Skopje in its name dispute with Athens. This is especially evident at a time when the Greek government is desperate to maintain social stability while pursuing austerity measures to reduce the country's mammoth budget deficit.

Second, the VMRO-led government made several provocative decisions designed to reinforce claims to an ancient regional identity that raised the temperature with Athens and reinforced Greek intransigence. For instance, the decision in 2007 to rename the airport in Skopje after Alexander the Great, the ancient king of pre-Slavic Macedonia, seemed calculated to provoke Greek sensitivities over ancient Macedonia's Hellenic heritage. By blocking the country's NATO and EU entry, Greece appeared to contravene its undertaking in the 1995 Interim accord not to allow the name issue to stand in the way of the country's membership in international organizations. Athens countered that Skopje had broken its own pledges by usurping the heritage of Ancient Macedon and implicitly making claims to Greek territory through symbolic gestures and educational textbooks.

Third, the long-term failure to find a renaming solution for Macedonia could negatively affect the Albanian coalition partner, the DUI, especially as the key foreign policy priority for all prominent Albanian leaders in the Balkans is to join NATO and move closer to the United States. Macedonia's stalled NATO accession is a source of frustration for Albanian representatives as it could lead to isolationism and nationalism in which the Albanians will be left stranded or even the targets of ethnic conflict.

Skopje's obstructed progress toward NATO and the EU will increase opportunities for political disputes to assume ethnic dimensions. The VMRO government could find itself facing a spiral of instability in which the Albanian position hardens and increasingly criticizes Skopje for failing to devise a solution with Greece. The VMRO-DUI coalition could dissolve, and Albanian party leaders may push for decentralizing, confederalizing, or even fracturing the state, a scenario that would send political shockwaves throughout the region. Albanian parties may begin to compete with each other—not for a share of government portfolios with the Slavic Macedonian parties, but over programs for autonomy or even separation and union with Kosova or Albania. The danger of a unilateral partition of Bosnia-Herzegovina and Kosova, coupled with Macedonia's national isolation and its poor economic prospects, would also encourage such internal destabilization.

A fourth negative scenario could witness Macedonia devolving into another exploitable "frozen state" for the Russian authorities in South Eastern Europe, alongside Kosova and Bosnia-Herzegovina. Moscow calculates that a Macedonian state that remains outside of NATO and the EU will become a growing source of dispute and even conflict that can preoccupy Washington and Brussels. This can enable Russia to expand its geopolitical agenda in Europe's east aimed at rolling back U.S. influence and neutralizing NATO's security functions.

Disputes with Greece.

Opinion polls in Macedonia indicate strong support for retaining the country's current name and dismissing any compromises with Greece. Some observers and officials have suggested a geographical qualifier for the country such as "North Macedonia" or "Vardar Macedonia," and there has been limited support in Athens for such options. However, the hard-pressed Greek government, in the midst of an economic crisis, may be less inclined to accept "Macedonia" in any part of the name of its northern neighbor in the future.

The VMRO government continues to tap into sentiments of national pride and needs Athens as a counterpoint to its national agenda. It has used the emotional issue of the state name and national identity to garner public support on the premise that it is defending the distinctiveness of the Macedonian nation. In this quest, history has been revised and manipulated by government officials to depict the current Slavic-speaking Macedonians as the direct genetic descendants of Ancient Macedonians, mirroring the historical nationalism in Greek policy.[153]

Any alterations to the country's name agreed to by Skopje can be exploited politically by government opponents and could even lead to conflicts between Slavic Macedonians. Officials in Skopje claim that about 70 percent of the country's population opposes any change in the country's name, and over 80 percent of Slavic Macedonians are against any adjustments. In the event of a referendum that rejects modifying the country's name and in effect disqualifies Macedonia from NATO and the EU, conflicts with Albanians are likely to intensify. Although the nationalist vote ensures VMRO a majority among Slavic Macedonians, in the long term, it undermines its attractiveness as a coalition partner for the Albanian parties. Some observers believe that Prime Minister Nikola Gruevski is less interested in NATO and EU membership than in staying in power by using to his political advantage the issue of unfair exclusion by Greece.[154]

The Greek government is mired in internal economic crisis and growing social unrest and can ill afford to be seen to surrender any element of Hellenic heritage over Macedonia. Facing even more stringent austerity measures demanded by international lenders to keep Greece solvent, Prime Minister Antonis Samaras has compared the country's predicament to the Great Depression in the United States following the Wall Street crash in October 1929.[155] As a consequence, future governments could prove more nationalistic and less willing to compromise with Skopje than its predecessors. Some analysts and officials in Macedonia believe that the Greek authorities are deliberately biding their time on the name agreement, calculating that Macedonia's position will be steadily weakened through internal instability and international exclusion.[156]

Some Macedonian observers speculate that Athens is actually promoting insecurity in the country by encouraging and exploiting ethnic and religious cleavages to achieve its objective of obliterating the Macedonian name altogether from its northern neighbor.

The simmering dispute has also boiled over in periodic incidents with Greece that further sour relations between the two countries.

In June 2012, Greek border services started covering the letters "MK" on Macedonian car number plates with a FYROM sticker.[157] Macedonia introduced new number plates in February 2012, saying they were needed to meet EU standards. The Greek move will have a negative impact on tourism, as Greece is a leading summer destination for many Macedonians, and drivers may object to having border officers tamper with their cars. It will also feed into nationalist polarization on both sides of the border.

Prime Minister Gruevski benefited from the visit of UN Secretary-General Ban Ki-moon to Skopje on July 24, 2012, by accusing Athens of deliberately stalling negotiations on the name dispute and blocking Macedonia's entry into NATO.[158] Gruevski also told Ban Ki-moon to urge Greece to respect the rights of the Macedonian minority in Greece, whose existence Athens does not even recognize. Raising the Slavic Macedonian minority question with Athens is guaranteed to exacerbate conflict and close more doors to further negotiations.

The number of people speaking Slavic Macedonian dialects in Greece has been estimated at between 10,000 and 250,000. However, the majority do not openly exhibit an ethnic Macedonian national consciousness and identity themselves as Greek. The government in Skopje claims that their avoidance of Macedonian identification is primarily due to fears of discrimination and repression by Greek officialdom.

Minority Frustrations.

Following the Albanian insurgency in northwestern Macedonia during the summer of 2001, the Ohrid Framework Agreement was brokered between Albanian guerrillas and the Macedonia government under the supervision of U.S. and EU mediators.[159] Much of the agreement has been implemented, including cultural autonomy, proportional Albanian representation in state institutions, and use of the Albanian language in municipalities where Albanians constitute 20 percent or more of the population.

Nonetheless, several factors have undermined Albanian commitments to the Macedonian state. These have included the rise of Slavic Macedonian nationalism inflamed by the name dispute with Greece, state capture by the VMRO party, an ethnically assertive Prime Minister who has headed the government since August 2006, the growing prominence of the Macedonian Orthodox Church, a decline in the independence of the media and judiciary, the limited number of Albanians in senior positions in government institutions and public enterprises, and setbacks in administrative decentralization and Albanian language use. Within the framework of the Ohrid agreement, there has been progress on implementing the law on languages, on decentralization, and on more equitable minority representation. However, continued efforts are needed in various domains, such as providing education in native languages and strengthening political dialogue.

The rights given to Albanians under the Ohrid Accords have also led to dissatisfaction among the Slavic Macedonian majority, especially among those who feel a loss of privileged status in the state sector amidst charges that merit has been sacrificed for ethnic quotas. This has caused resentment against Albanians and latent opposition to power sharing and the redistribution of public resources. In the event of economic stress, such resentments can be politically manipulated to fuel inter-ethnic disputes. According to some surveys, more than two-fifths of Macedonians believe there is a high risk of violent ethnic conflict. Their views are shared by a slightly lower number of Albanians in Macedonia.[160]

The "national renaissance" campaign and costly urban renewal program (Skopje 2014) pursued by the VMRO administration is focused on asserting Macedonian identity and developing an ancient heritage that largely neglects Albanians and other minorities. The attempt to depict the current Slavic-speaking Macedonians as direct descendants of ancient Macedonians has magnified conflicts with Greece and contributed to excluding and alienating the Albanian population, which is characterized as outsiders rather than the claimed descendants of ancient Illyrians who inhabited parts of the Macedonian region since before 1000 BC. The nationalist focus on identity and deepening divisions between "patriots" and "traitors" will contribute to exacerbating inter-ethnic tensions at a time of economic uncertainty.

If the "national renaissance" or "antiquization" program of state-promoted nation building is pursued at the cost of international integration, it will delegitimize the Macedonian state among Albanians and increase demands for federalization and bilingualism. As both propositions remain unacceptable to the Macedonian majority, this will provide another recipe for conflict. The opposition Democratic Party of Albanians (DPA) has already called for "nonterritorial federalization" and a bicameral legislature to accommodate Albanian political aspirations.[161]

According to the EU Commission, Macedonia has continued with reforms related to its EU accession bid, although several challenges remain.[162] The parliamentary elections on June 5, 2011, were generally in line with international standards. Progress has been made in the fields of the judiciary and public administration, notably with regard to the legal framework. However, problems are also evident in such areas as freedom of expression in the media, judicial and administrative reform, and combating pervasive corruption. Additionally, the absence of constructive dialogue between government and opposition elements weakens the functioning of state institutions.

Critics charge the VMRO administration with state capture, combined with autocratic and populist politics that are promoting instability and diminishing the country's chances for EU and NATO accession. Indeed, EU officials have expressed worries that Macedonia is backsliding on reform in the face of Greece's blockade of the country's EU membership talks. Macedonia gained EU candidate country status on December 17, 2005; and since 2009, the EU Commission has recommended a start to accession talks but failed to offer an actual date.[163]

In an indication of growing frustration among Macedonia's Albanians, on October 14, 2011, the national census was abruptly terminated following the resignation of the State Census Commission, which protested against various counting irregularities.[164] The previous census, held in 2002, recorded Albanians at 25 percent of a total population of 2,022,000. The figure is dismissed by all Albanian leaders, who claim that the actual number exceeds 30 percent of the population, while some Slavic Macedonian spokesmen maintain that the figure is under 20 percent.

Population numbers have important ramifications for such questions as language use and representation in state institutions. In the 2011 census, Albanian members of the Census Commission demanded that expatriate citizens who visit Macedonia at least once a year should also be counted. This was rejected by the Macedonian members. Moreover, there was a broader lack of clarity regarding counting methodology, and the entire exercise became heavily politicized.

The economic recession and high unemployment, estimated at 31 percent by the close of 2011, has exacerbated public distrust in state institutions and in the governing coalition. The state debt is growing, while economic growth is stagnant and lagged behind most countries in the region. The resources spent on renovating Skopje and creating an ancient heritage have led to charges by Albanian leaders that only limited resources have been invested for the benefit of the Albanian community and disparities in living standards with Slavic Macedonians are rising. If these negative trends are prolonged, they can precipitate social unrest and rekindle latent inter-ethnic tensions. Political leaders in both the Slavic and Albanian communities may divert dissatisfaction with their own performance into communal confrontations to masquerade as the protectors of distinct ethno-national interests.

Ethnic Escalation.

Ethnic tensions in Macedonia are not confined to relations between Slavic and Albanian Macedonians but have involved Slavic Muslims and Turkish populations. In February 2012, OSCE High Commissioner on National Minorities Knut Vollebaek, who was visiting Skopje, stated that the upsurge in ethnic tensions was a wakeup call for the country to make greater efforts to rebuild community trust.[165]

On January 13, 2012, Macedonia's Muslim communities expressed their indignation over a carnival in which Orthodox Christian men mocked Muslims by dressing as women wearing burqas.[166] The Vevcani carnival attracts thousands of visitors, and local residents traditionally wear elaborate masks. The most common costumes include devils and demons. The festival incident prompted violent demonstrations by Muslims, who accused members of the majority population of stoking hatred against them. Some protesters attacked buses and defaced a Macedonian flag in the town of Struga and replaced it with a green flag to represent Islam. The Saint Nikola church in Labuniste, a village near Struga, was partially burned by unknown perpetrators at the end of January 2012 as tensions between Christians and Muslims accelerated.[167]

Macedonian Muslim leaders called for restraint while accusing the government of promoting Islamophobia through the organs of official propaganda. The head of the Islamic Community demanded an apology from the mayor of Vevcani, a Slavic Macedonian village in an Albanian majority region. Following a number of violent incidents, the national Commission for Religious Communities held an emergency session. Some locals claimed that the incidents were politically motivated, evidently to sow divisions between Christian and Muslim Macedonians and between Slavs and Albanians. The incidents in Macedonia also had reverberations in Kosova, where the Macedonian embassy building in Prishtina was attacked during a protest organized by radical groups.[168]

In a further indication that ethnic relations remained tense and liable to provocations, the shooting of five unarmed young Slavic Macedonians near Skopje on April 12, 2012, raised fears of revenge attacks on Albanian communities even though there was no evidence that Albanians were involved in the shootings.[169] On May 4, 2012, several thousand Albanians protested in Skopje against the arrest of Muslims allegedly implicated in the murders. Police arrested 20 Albanians during an operation in several villages around the capital, and five were subsequently detained on terrorism and murder charges.[170]

Government officials claimed that radical Islamists were using the arrests as an excuse to whip up tensions. Shukri Alia, blacklisted by the EU and sought by the Macedonian police for murder and armed attacks on two Skopje police stations, is reportedly leading the efforts to organize protests. Police believe he is hiding in Kosova and planning to provoke a civil war by pushing for the secession of western Macedonia.[171]

According to Macedonia's security experts, an estimated 5,000 battle hardened nationalists are in the country, including some radical Islamists from the wars in Bosnia-Herzegovina, Kosova, and Macedonia. Other analysts believe that the threat is exaggerated and used as a smokescreen for anti-Albanian and anti-Muslim militancy. Radical Islam has a toehold in Macedonia through the Salafist and other ultra-conservative movements, and Muslim identity may figure more prominently where there is conflict with the government over its close ties with the Macedonian Orthodox Church. Any favoritism shown to the Orthodox Church serves to strengthen an Islamic identity among Slavic, Turkic, and Albanian Muslims. Additionally, radical Islamists are seeking inroads through charitable, humanitarian, and educational work among the poorest sectors of society. This could constitute a long-term danger to moderate Islamic traditions and to interconfessional tolerance.

Various theories have been offered regarding the cause and context of the post-carnival protests and violence, and to what degree these were spontaneous or pre-planned.[172] These have included alleged collusion between the Greek government and local Albanian radicals to destabilize Macedonia; attempts by pan-Albanian nationalists from Albania and Kosova to provoke intercommunal disputes; conflicts between Albanians and Slavic Macedonian Muslims instigated by the latter's allegations that they are pressured to declare themselves as Albanians; disputes within the Slavic Muslim community for political influence and greater rights commensurate with the post-Ohrid Albanians; attempts to demonstrate the existence of a distinct ethnicity, the Torbeshi, who consider themselves neither Bosniak nor Macedonian Muslims while demanding distinct political benefits and institutional representation; and a foreign terrorist threat promoted by Tablighi Jamaat, the Islamist missionary group active among various Balkan Muslim populations.

Some local analysts believe that the clashes may have been engineered by the VMRO administration to test the waters for future conflict and raise nationalist sentiments from which it will benefit among Slavic Macedonians. Many Macedonians resent the Ohrid peace accord because they feel Albanians have gained special privileges. They view the Albanians as a fifth column planning to divide Macedonia and merge territories with Albania and Kosova. There is also rising resentment that Slavic Macedonians in Greece and Bulgaria have not been accorded commensurate rights with Albanians in Macedonia, and indeed that Macedonian ethnic identity has been denied by Athens and Sofia. Bulgarian officials in turn have claimed that Bulgarians in Macedonia face official discrimination and persistent human rights violations.[173]

Inter-ethnic incidents continued throughout 2012, including gang attacks on Albanian and Macedonian students and the shooting of two young Albanians in Gostivar by an off-duty Macedonian policeman. This led to a demonstration on March 1, 2012, by some 10,000 Albanians in Gostivar remonstrating against alleged police brutality.[174] Minister of the Interior Gordana Jankulovska asserted that radical forces wanted to disturb inter-ethnic relations. Protesting Albanians described the killings in Gostivar as ethnically motivated and accused the police of downplaying their significance. Sociology professor Hasan Jashari concluded that ethnic tensions were again on the rise and blamed poor political leadership, which allowed national identities to be strengthened at the expense of a civic identity.[175] In response to growing ethnic tensions, the emergence of ultra-nationalist Macedonian groups was reported throughout 2012 and included the Christian Organization (Hristianska Organizacija), which released a video showing members burning Albanian flags and chanting anti-Albanian slogans.[176]

Another source of friction has been over historical interpretations, especially over the longevity of habitation by different ethnic groups in the Macedonian region. In the spring of 2012, the Macedonian and Albanian academies of arts and sciences sought to settle such disputes, but government officials charged that they were interfering in Macedonia's internal affairs.[177] One of the key issues to be addressed concerned the

Macedonian encyclopedia. Albanian historians dispute the text because it claims that Albanians were newcomers to the Balkans and not native to Macedonia. The encyclopedia describes Albanians as "mountain people" who descended into Macedonia during the Ottoman conquest in the 15th and 16th centuries. The Albanian Academy considers this a skewed nationalist interpretation that blatantly disregards the Illyrian roots of contemporary Albanians.

Meanwhile, some historians in Skopje claim that Macedonians are not recognized as a distinct and ancient people in Albanian text books but as "Slavomacedonians," thus indicating that they are relative newcomers to the Balkans, dating to the Slavic migrations in the 5th and 6th centuries AD. Neither side seems to recognize that ethnogenesis and nation building involve a long and complex process, and that few, if any, modern European nations can claim direct genetic or cultural ancestry from any distinct ancient population. This is due to centuries of mixture and assimilation between older and newer populations. If language is a relatively constant variable in an otherwise shifting structure of group identity, then it is especially difficult for present-day Macedonians to claim a pure pre-Slavic heritage. Nevertheless, contrasting and conflictive interpretations about historical lineages and territorial longevity impregnate contemporary politics and intercommunal relations. Identity, territory, and statehood will continue to be contested, and such disputes may increasingly resonate among a discontented public.

VI. INTERNATIONAL DEFICIENCIES

For over 20 years, international actors have been involved with the Western Balkans in a multitude of roles: as diplomatic mediators, humanitarian agencies, peacekeepers, combat forces, democracy promoters, and state builders. In their most recent incarnation, international institutions are engaged in a policy of regional integration into the EU and NATO. However, the process of accession has taken much longer than initially expected at the end of the Yugoslav wars. As a result, several Western Balkan states are in danger of being left out of both institutions indefinitely, thus contributing to uncertainties about domestic reform and regional stability.

The EU is mired in its own internal economic crisis, which will have repercussions for social stability, political decisionmaking, and institutional development, and with profound reverberations for further Union enlargement. In addition, the United States has been reducing its military and political presence in the Western Balkans on the assumption that the EU would complete the task of international inclusion and permanent regional security with Washington's secondary support. The danger persists that insufficient impetus from Brussels and an inattentive or diplomatically disengaged America may allow for the mushrooming of conflicts that could increasingly destabilize the region.

European Stagnation.

The EU is wracked by internal confusion and indecision, which will have a direct impact on Brussels' policy toward the Western Balkans and the membership prospects of each state. While the immediate concern is the future of the Monetary Union—the feasibility of a fiscal union, and eventually a political union—the ultimate danger is runaway debt that can generate a deeper economic and social crisis. In stark figures, government debt as percentage of GDP has ballooned to unmanageable proportions in several countries. In Greece, the figure stands at 165 percent; in Italy, 120 percent; in Portugal, 108 percent; and in Spain, almost 70 percent and growing.[178] Meanwhile, the Union as a whole appears to be heading toward prolonged recession, as even the larger economies remain stagnant.

On June 18-19, 2012, EU leaders agreed at the G20 Summit in Mexico to construct a more integrated banking system. This was an effort to stem a debt crisis that threatens the survival of the euro and undermines the global economy.[179] Although Germany and several Eurozone partners laid out concrete steps toward financial integration, EU President Jose Manuel Barroso asserted that it would take time for the 17 Eurozone capitals to agree on establishing a fiscal union. Moreover, countries that are fearful of losing their sovereignty will resist such measures.

German Chancellor Angela Merkel has been supportive of closer fiscal integration that would involve ceding sovereignty over national budgets to a central authority in Brussels. Other leaders, including French President Francois Hollande, have doubts about transferring fiscal powers but support the issuing of Euro bonds that would involve a sharing of debts—a plan that Berlin opposes. Meanwhile, financial markets are desperate for an EU roadmap leading to closer fiscal, banking, and political integration necessary to make the single currency a viable long-term proposition. Without assurances that the Eurozone is planning to "mutualize" the debt owed by all members, bond investors will accelerate their departure from the struggling Mediterranean economies.

As European governments ponder, the financial crisis is deepening, economic performance is deteriorating, and the common currency is in danger of unraveling. In addition to the Greek economic meltdown, investors and depositors have lost faith in other weak economies, especially those of Italy, Spain, and Portugal. This has raised anxieties about government debt defaults amid preparations for rescue packages from the European Stability Mechanism (ESM), established in 2011 to shield heavily indebted governments but whose funds remain limited.

The majority of voters in Germany and most West European states oppose further EU bailouts for Greece or other Mediterranean economies, as they feel they are carrying too much of the financial burden. However, economic stagnation in the larger Mediterranean economies such as Spain and Italy will also hurt Germany, as it will scale back demand for German exports on which Berlin's fiscal strength depends. In addition, demand for EU imports is weakening global markets. If the European-wide recession deepens, north European leaders will experience more problems in acquiring the finances necessary to contribute to fiscal lifelines to southern Europe. Conversely, accelerated borrowing will only buy limited time for confronting the spiraling debt.

Deep spending cuts and austerity measures may be the only way to avert even more catastrophic debt that will be passed on to the next generation and necessitate huge tax hikes, drastic cuts in public expenditure, and lower living standards.[180] The financial crisis has resulted in increased borrowing, especially in the south European economies, and the accumulation of massive debts to stimulate economies out of recession. Unfortunately, this may be a short-term palliative disguising a long-term fiscal disaster.

EU governments are also curtailing spending to stymie ballooning national debts, but deeper austerity will raise unemployment levels across the Eurozone. The International Labor Organization, in its 2012 report on global trends, warned about mass unemployment because of cuts in government spending. It forecasts that 4.5 million more jobs could be lost in the Eurozone over the next 4 years and that the total number of jobless will reach 22 million.[181]

A growing army of unemployed, combined with unsustainable government debt, will fuel social unrest and political turmoil in the years ahead. Deficit spending cannot continue indefinitely where debts exceed productivity, borrowing becomes prohibitive, defaults are threatened, and investors abandon unprofitable countries. To become competitive in an increasingly complex global market and to ensure steady economic growth, the EU needs to stimulate business. The southern part of Europe, in particular, needs to undergo a business revolution to emerge from austerity and restore economic confidence. Politicians need to dispel the false dichotomy between austerity and growth, as the most sustainable solution necessitates both budgetary discipline and the stimulation of private enterprise.

Europe's financial crisis and economic underperformance have contributed to the economic downturn in the Western Balkans by diminishing direct foreign investment, international aid, and diaspora remittances. Economic prospects for the region as a whole remain uncertain. Financial sector vulnerabilities constitute the biggest risk, because much of the banking system is foreign owned and most countries are reliant on funding from abroad. Economic activity in the region continued to weaken throughout 2012.[182]

The Greek crisis is also having a direct impact on several Western Balkan banks, which are either Greek-owned or exposed to Greek debt. If they collapse, this will have a visible impact on business and the availability of credit in the region. If the EU's economic malaise, driven by the sovereign debt crisis and the necessity of protracted governmental austerity measures, persists, this will usher in an era of economic stagnation in the Western Balkans as fewer resources will be available for non-EU countries.

EU Leadership Deficit.

With Europe's sovereign debt crisis accelerating, fingers point at the absence of EU leadership. No single official in the beleaguered bloc exhibits the charisma and courage necessary to push through difficult decisions, and no official benefits from a broad international consensus. The EU has created a thick web of structures and authorities to create an impression of effective leadership, but without a credible political union, none of these carry significant weight. President of the European Council Herman Van Rompuy, President of the European Commission José Manuel Barroso, and President of the European Parliament Martin Schultz each have limited responsibilities, and two of them have not even been democratically elected.

Neither Rompuy nor Barroso are in a position to determine key EU decisions, such as the proposed creation of a fiscal union or tighter political integration, and neither will challenge the leadership in Berlin or Paris. However, Germany and France no longer have leaders with the stature of Konrad Adenauer or Charles de Gaulle, but instead have politicians whose chief concern is re-election. Additionally, the faltering performance of the French economy has dissipated the voice of Paris in European affairs, while German Chancellor Angela Merkel is wary of turning the EU into a political union, lest this is perceived as a cloak for German hegemony.

The lack of leadership has been evident in foreign policy, which was supposed to become coherent under the 2009 Lisbon Treaty, and in EU internal policy. Union leaders have acted slowly and inadequately in response to the escalating financial crisis among several Mediterranean members. As a result, the EU's monetary union may be fragmenting faster than policymakers can pour money in to repair it.

Eurozone leaders recently agreed to establish a joint banking supervisor for the single currency area based on the European Central Bank. This is envisioned as a first step toward a European banking union to prevent bank runs or collapses that send shock waves around the continent. The Eurozone's permanent bailout fund, the ESM, containing 500 billion euro, could inject capital directly into banks once the joint supervisor is established. However, critics charge that this move is too little, too late. In actuality, the Union is dividing between the northern creditor countries such as Germany and Holland, whose borrowing costs are at an all-time low, and southern debtor countries such as Spain and Italy, who find it difficult to borrow money to service their ballooning debts.

With EU leadership adrift, the wealth gap between north and south is accelerating, and even large bailouts from the ESM will only delay the crisis for a few months. Given the current trajectory, the collapse of the monetary union may well occur before the creation of the projected fiscal or banking unions. Some analysts claim that the looming collapse of the Monetary Union will finally force its leaders to push for a political union and a joint European constitution that would centralize decisionmaking. But given the prevarications and half-measures witnessed over recent years, the Union is more likely to disintegrate before it becomes an actual Union.

Some analysts contend that for the euro to survive, Eurozone members need to establish federal institutions with a common budget and undertake deep economic reforms to ensure a rate of growth sufficient to decrease public debt. Throughout this transformation, defense spending would diminish even further, and the EU's Common Security and Defense Policy (CSDP) would lose its remaining relevance. These factors would precipitate the wholesale decline of the EU and encourage the emergence of a core group increasingly divorced from other member states and from non-EU aspirants.

Even if the EU survives, it would be in a weak position to act as a security provider or an effective deterrent of conflict. As a result, the security situation could deteriorate in some sub-European regions, including the Western Balkans, which could be "left in a limbo between Russia and Turkey on the one hand, and an Asia-focused U.S. and a federalizing European core on the other."[183]

Nationalist Specters.

Economic recession has led to growing feelings of economic and social insecurity that inflame protest movements and more radical political options. Extensive austerity measures can lead to social conflicts that will further handicap EU reform. Opposition to neo-liberalism and globalism, which are currently perceived as worsening socio-economic inequalities, could favor protectionist and economic nationalist positions in several member states.

No one is certain whether a solution to the EU's financial crisis can be found within the existing institutional framework. According to one prominent French analyst, the combination of low growth and inadequate federal mechanisms in budgeting casts doubts on the Union's existence in its current form.[184] A possible collapse of the euro would undermine the single market and precipitate the disintegration of the EU itself. In a worst case disintegrative scenario, the EU would acquire populist, protectionist, and nationalist governments, and NATO and the EU would need to develop ways to manage crises within EU countries as well as possible conflicts among member states.[185]

For several decades, the pendulum between center-left and center-right has swung within relatively narrow confines. Two or three major parties monopolized governments in EU states, sometimes in coalition with smaller formations, but their policies and ideologies overlapped. This structure was stable when economic conditions for the majority of citizens were favorable. But in recent years, it has been unbalanced through growing economic uncertainty and grievances against mismanagement by the governing elite. While one election cycle may simply switch the center-right with the center-left or vice versa, two cycles of budget cuts and austerity measures can bring new players to the forefront.

Lurking in the political wings and seeking to benefit from public fear, confusion, and anger are three radical tendencies: the brown, the red, and the black.[186] Citizens disillusioned with the political establishment can veer toward these extremes. Radical nationalist and populist groupings have always existed on the political margins. They may acquire greater popularity as the prospect for Union disintegration looms on the horizon. Some are heralding the dawn of a "Europe of nations," free from the shackles of the Brussels bureaucracy.

The radical browns dismiss the centrist politicians as traitors to the nation, whether because they bow to Brussels or open doors to immigrants who allegedly steal jobs and welfare from local inhabitants. Anti-immigrant passions are combined with Islamophobia, racism, and anti-Semitism; and the brown patriots are depicted as protecting the nation from assimilation or extinction. Golden Dawn in Greece is the most successful example of this phenomenon; others are waiting for their opportunity in Italy, Spain, France, and elsewhere, and they could serve as examples or stimulators for ultra-nationalists in the Western Balkans.

Economic hardship has also raised the specter of the radical reds. They are less coherent than in their 20th century incarnation but share with the brown nationalists a disdain for globalization, which is often a shorthand for anti-capitalism and anti-Americanism. The third stream, the black nihilists, have no credible political or economic solutions, as their primary aim is to destroy the existing system. They tap into the frustrations of unemployed youth or romantic revolutionaries from the middle class. Although they do not participate in elections, their street protests, vandalism, and attacks on law enforcement contribute to radicalizing and polarizing the political atmosphere.

The three colors can converge during times of troubles, whether through joint actions or violent street clashes against each other. A more ominous scenario would materialize if they succeed in using the political process to capture seats in parliament or positions in local, regional, or national governments, or if some of their political prescriptions are adopted by mainstream parties in attempts to regain broader public support.

Weakening EU Magnetism.

In 1999, the European Commission initiated the Stability and Association Process (SAP) with the Western Balkan countries.[187] At the Feira Summit in June 2000, the EU explicitly offered all states in the region a credible prospect of membership once a series of conditions for qualification had been met. SAP was boosted in June 2003 when the Thessaloniki Summit specified the full range of necessary reforms for all Western Balkans states to become EU members. Since that time, the region has witnessed uneven progress toward accession and a waning EU appetite for further enlargement among citizens of member states.

Albania entered into an Stabilization and Association Agreement (SAA) with the EU on June 12, 2006, and applied for EU membership on April 28, 2009. It was officially recognized by the EU as a "potential candidate country" but was not awarded candidate status, as it failed to meet specific political recommendations prepared by the European Commission.[188] On October 10, 2012, the Commission's Progress Report recommended candidacy status for Albania once it complies with recommendations to complete key judicial and public administration reforms.

Serbia obtained an SAA on April 29, 2008, and applied for EU membership on December 22, 2009. The European Commission recommended making Serbia an official candidate on October 12, 2011, which was granted on March 2, 2012.[189] Bosnia-Herzegovina signed an SAA on June 16, 2008, which will enter into force once its ratification process has been completed.[190] Macedonia was the first country to sign the SAA on April 9, 2001, and has been a candidate for EU accession since December 2005. In 2008, the resolution of the name dispute with Greece was added to official preconditions for EU entry.[191] Montenegro signed an SAA in 2007 and officially applied to join the EU on December 15, 2008. It obtained official candidate status on December 17, 2010.[192]

The shortcomings displayed by Romania and Bulgaria since they entered the EU on January 1, 2007, in combating official corruption, ensuring judicial independence, and curtailing political influence in state institutions has led EU capitals to conclude that more extensive preparations by candidates were necessary before EU entry was permitted.[193] Hence, more stringent benchmarks are in place for the Western Balkan aspirants, and their entry is unlikely to be hastened by political or geostrategic calculations. Indeed, the pace of accession talks with candidates is unlikely to accelerate after Croatia's expected accession in the summer of 2013.

EU magnetism is weakening in the region. The premise that Union membership would become a source of irresistible attraction that would propel forward the reform process in every capital has not come to fruition. Euroscepticism has increased among politicians and publics in the Western Balkans. For many current office holders, EU membership would necessitate stricter compliance with standards of transparency and competition, thus threatening their positions and incomes. For many citizens, the pull of EU membership has diminished amidst doubts that the EU will quickly absorb new states. Both factors have tempered the appetite for domestic reform.[194]

Given the long drawn-out process even before accession talks actually begin, the prospect of EU membership alone does not automatically provide an impetus for structural reform or resolve lingering conflicts. For example, in Serbia, support for EU accession dropped from 76 percent in 2003 to 46 percent in October 2011, with 36 percent actually opposed to membership.[195] This is the result of skepticism over the likelihood of accession and a diminishing desirability for membership in the light of the EU's economic crisis. Although EU candidate status was achieved in March 2012, Belgrade did not receive a date for the start of accession talks, and the entry process is likely to stretch well beyond 2020.

Visa-free travel to the EU has been interpreted in the region not as a step toward membership but as a side-step to delay and prolong the accession process by offering attractive palliatives. The EU lifted visa requirements for citizens of Serbia, Macedonia, and Montenegro in December 2009 and for citizens of Albania and Bosnia-Herzegovina in November 2010.[196] In June 2012, Kosova received a roadmap for gaining EU visa liberalization, specifying conditions Prishtina needs to fulfill before its citizens can travel to Schengen states without visas. EU officials have warned that unless the Western Balkan governments reduce the high number of asylum seekers, visa liberalization could be suspended.[197] Schengen members may be allowed to restore normal border controls for a period of 6 months with the possibility of extending the measure for a further 6 months. Any reversal of the visa liberalization program will in turn undermine the Union's credibility throughout the region.

"Accession fatigue" or "reform fatigue" is becoming more noticeable in the region, and EU membership may be viewed increasingly as a mirage. In the case of Macedonia, disillusionment is linked with blocked EU accession prospects because of the name dispute with Greece. In opinion polls conducted during 2011, 72 percent of Macedonians expressed a lack of trust in EU policy toward the country.[198] Similar feelings are likely to grow in Kosova if the country's prospects toward EU accession are blocked by the five EU states that do not recognize its independence.

Another criticism of the EU is that certain initiatives supported by Brussels failed to have a significant regional impact. For instance, one of the objectives of the SAP was to enhance regional cooperation especially in the economic domain.[199] Despite the forging of bilateral free trade agreements and the extension of the CEFTA in 2006 to encompass the Western Balkans, the economic impact has been marginal. CEFTA is seen as secondary to EU accession and has added little competitive advantage in attracting foreign investment.

Western Balkan leaders and publics harbor resentments against EU conditionality, prolonged or indefinite timelines, and unclear signals on membership criteria. This has created the perception of receding accession targets, as numerous additional conditions are required and a sense that the EU is seeking to indefinitely postpone the process. For example, Union representatives have adopted an ambiguous position on the necessary modifications to the Bosnian constitution to bring it in line with EU standards.[200] While European Commission President Jose Manuel Barroso stated that constitutional reform was not a condition for signing the SAA with the Union, EU Enlargement Commissioner Olli Rehn stressed that such reform was vitally important for Bosnia-Herzegovina to make progress toward EU membership.[201] The EU appeared to lack consensus about its norms for state building and membership conditionality.

Such confusion has contributed to the evasion or even reversal of reform programs and accentuated the notion that EU conditionality is flexible and negotiable and lacks a timetable and a clear list of state-building targets. Moreover, if EU conditionality is intended to create a functioning state, political leaders who are preparing partition will not comply with such programs. Indeed, Serbian political leaders in Bosnia-Herzegovina do not view the benefits of EU accession and the construction of a more functional central state as outweighing the costs of abandoning RS autonomy and its attendant political powers and economic benefits.[202]

The EU has failed to demonstrate its effectiveness in helping Bosnia-Herzegovina move toward EU candidacy and to resolve deep-rooted ethno-political disputes. It has been unable to buttress a stronger and more functional state government, which is essential for purposes of EU integration. This fits with broader conclusions that the EU state-building model has failed to live up to expectations in the Western Balkans and has inadequately addressed the question of ethno-national representation in state institutions.[203] EU members possess a diversity of state structures, from highly centralized states, which do not recognize minority rights, such as France and Greece, to decentralized federations, including Germany, Spain, and the United Kingdom. As a result, there is no single standard that could be applied to all Western Balkan countries.

Europe's Softening Power.

The EU has prided itself on its "soft power" attributes that enabled it to exert significant influence over aspirant countries. However, the impact of the accession weapon is weakening the EU's "soft power" arsenal. For instance, Bosnia-Herzegovina signed an SAA with the EU in June 2008, but since then Sarajevo has failed to implement the reforms necessary to move forward in the integration process, and the EU seems largely powerless to encourage faster progress. There has been no implementable agreement between political leaders on three key EU conditions: passing a law on state aid, organizing a national census, and amending the constitution to preclude discrimination for elected office on ethnic grounds, in line with the ruling of the ECHR.

Divisions among EU states on recognizing Kosova's independence, with five countries abstaining, have had a negative impact on the effectiveness of the EULEX mission and sparked disputes whether the EU representation in Prishtina should be called a liaison mission or an office. Moreover, as the EULEX presence is "status neutral," it is not mandated to play a role in state building and ensuring progress toward Union integration.

EULEX has also failed to establish its mission in the north of the country, to control the northern border with Serbia, to prevent smuggling, or to convict and imprison criminal kingpins and corrupt officials. Kosova is the only country that does not have contractual relations with the EU because five EU member states have not recognized its statehood. Four of the five states are also NATO members; hence, Kosova is also obstructed from NATO integration. Additionally, the EU has no jurisdiction over border demarcations and has not developed applicable rules or procedures, even though regional cooperation is supposed to be an essential part of qualifying for EU integration.[204]

The EULEX mission has proved disappointing to many Kosova citizens and is perceived as failing to fulfill its mandate to build a strong system of laws to combat corruption and organized crime.[205] This has seriously dented public trust in EU institutions. Even the dialogue initiated by EU mediators between Belgrade and Prishtina has had mixed results. Discussions between Belgrade and Prishtina on Kosova's representation in regional organizations ended in agreement on February 24, 2012.[206] But they also unleashed a storm of controversy among Kosovar Albanians regarding amendments to the country's name.

For Kosova, which currently does not have membership prospects, EU conditionality is irrelevant, and the Union lacks the leverage to reward compliance or penalize noncompliance. As one analyst from the region concluded, "either the EU will devise a bold strategy for accession that encompasses all Balkan countries as new member states, or it will become mired as a neo-colonial power in places such as Bosnia and Kosova."[207]

There is a core disagreement on the nature of international involvement in the Western Balkans between proponents of "guided reform" and of "local ownership." The debate on the degree of involvement by outside actors may actually contribute to paralyzing decisionmaking. It has also confused citizens in the region, who veer between supporting a greater EU role to push through necessary domestic reforms and rooting out local corruption and resentment against overbearing international interference in domestic affairs.

The proponents of "local ownership" and a less obtrusive outside role argue that effective state building can only be accomplished when international intervention is reduced and local communities have a commitment to national institutions. In contrast, the interventionists argue that allowing local politicians to determine the future of the state could result in disputes that lead to government paralysis, territorial division, and regional conflict that would in turn necessitate an even more intrusive international role.

In practice, it is difficult for international actors to assist in constructing a functioning and stable state where there is insufficient domestic consensus among political leaders with regard to its structure and functions, as in Bosnia-Herzegovina.[208] On the other hand, any existing consensus would be seriously challenged if international instruments are weakened or key international players disengage from the region. Maintaining post-conflict stability was understandable in the first few years after the Yugoslav wars. However, it has become increasingly tied to maintaining a political status quo and tolerating inadequate or dysfunctional governments.

Such short-term stability may not engender long-term security as corruption and mismanagement among governing elites, in combination with their personal and national ambitions, undermines economic development, raises social tensions, delegitimizes state institutions, and disqualifies these countries from EU entry and leaves them exposed to renewed conflicts.

U.S. and NATO Downsizing.

There is a pervasive feeling in the Western Balkans that without the presence of American troops, regardless of the precise number of ground forces, stability cannot be guaranteed if the simmering conflicts over statehood or political representation were to escalate. There is also a growing sense that U.S. interest, involvement, and effectiveness are dissipating because Washington is focused on more pressing national security questions. As a result, the U.S. ability to control or manage developments in the region may be diminishing.[209] The absence of European military deterrents and decreasing U.S. engagement could encourage nationalist and revisionist forces to raise their profile and provoke fresh conflicts in the years ahead.

Kosova presents a pertinent example of the challenges of military downsizing in an uncertain political and economic climate within a disputed state.[210] Since 2000, NATO has consistently reduced its KFOR troop presence in the country. In the immediate aftermath of NATO's bombing campaign in June 1999, KFOR deployed around 60,000 troops in the country. This total was periodically reduced; in October 2010, the North Atlantic Council announced plans to streamline KFOR's presence from under 10,000 troops to 5,500.

The United States deployed 7,000 troops in Kosova in June 1999, and the total shrank to approximately 2,500 by 2004. By early 2012, there were 1,447 U.S. Soldiers in Kosova stationed at Camp Bondsteel and Camp Nothing Hill near Leposavić.[211] Despite their modest current number, the presence of U.S. troops is considered by the Kosova government and citizens as an invaluable deterrent to new outbreaks of violence. NATO forces also continue to perform an important role in de-escalating potential conflicts in northern Kosova and demonstrating to outside powers that the Alliance is committed to regional security.

At least three challenges must be considered when assessing the security situation in Kosova during NATO's downsizing: the situation in northern Kosova; relations between Kosova and Serbia; and the development of Kosova's own security capacities in assuming NATO responsibilities. Although the main security challenges are unresolved political issues, economic and social factors also generate tensions and could contribute to violent conflict. KFOR's troop reduction in Kosova should be conducted in relation to the capacities of local institutions to assume responsibility for a secure environment. In this regard, it is necessary to increase the professional capacities of the Kosova Security Force (KSF) to assume appropriate responsibilities.

Northern Kosova presents a serious obstacle to security, as over 60 percent of the border line is not under the control of the Prishtina authorities. If account is taken of further NATO troop reductions, the government must increase its level of responsibility in border control. The KSF was established within a fixed mandate as a civilian protections force. To date, its competencies only entail crises response, including fire fighting, handling hazardous material, search and rescue, demining, and disaster relief. Additionally, Kosova's police need to develop cooperation with international organizations, such as the European Police Office (EUROPOL) and European Agency for the Management of Operational Cooperation at the External Borders of the Member States of the European Union (FRONTEX), which are responsible for preventing and combating terrorism, drug trafficking, and organized crime.

After 9/11 and Washington's preoccupation with military operations and political missions outside of Europe, there was a prevailing realization that the United States would gradually withdraw its forces from Bosnia-Herzegovina and leave the country to EU supervision en route to Union integration. The initial U.S. military presence in NATO's 54,000 strong Implementation Force in December 1995 numbered 15,000. This total shrank to about 900 by 2004 and then to a handful of military personnel by 2012.[212] The NATO operation was replaced by the smaller EUFOR Althea in December 2004, consisting of approximately 2,500 troops from EU member states. The number was subsequently reduced to about 1,300, although EU officials claim that they can be augmented by Over the Horizon Forces consisting of up to four battalions.[213]

Meanwhile, the OHR was rendered powerless after 2006, and the Bonn Powers that enabled a more direct role for the OHR in pushing through reforms were largely discarded. Since 2002, the High Representative also serves as the EU Special Representative (EUSR) to Bosnia. All of the High Representatives have been from EU countries, while their principal deputies have been Americans. Several EU countries would prefer to see the OHR closed, thus further undercutting the American role, because Washington needs the OHR to operate legally in Bosnia. Washington has sought to preserve the OHR and to maintain the Bonn Powers that allow for greater political intervention in cases where the principles of the Dayton Accords are undermined by Bosnia's leaders. Bosnian politicians are aware that U.S. involvement is declining and its leverage is decreasing, while the EU without American leadership is notorious for its slow motion and national divisions. This can embolden political forces seeking greater decentralization or outright separation.

If other security priorities and prolonged budget cuts result in the United States withdrawing from several NATO missions, the Western Balkans may pose one of the immediate tests.[214] The U.S. President may find it difficult to make the case for renewed American involvement even if security in the region begins to unravel. The EU or the European pillar of NATO would be expected to handle any local crisis without any significant American military engagement, especially of U.S. ground forces, in a region no longer viewed as a priority for America's national interests. This would also place new stresses on the NATO alliance and herald potential disputes between European capitals over the merits and methods of military intervention.

In the event of violent clashes in Bosnia-Herzegovina, EUFOR would be dependent on a considerable external backup of European troops. An additional complication would be simultaneous conflicts in different states that prevent military forces from responding adequately in each scenario. According to local analysts, EUFOR's current configuration, strength, deployments, and posture place it in danger of failing even under moderate strain.[215] It has lost its ability to provide a credible deterrent, while its reactive capacity may be insufficient to defuse an escalating conflict.

Given the persistent political tensions in Bosnia-Herzegovina, EUFOR may be increasingly called upon to maintain or restore a safe and secure environment. However, without the political will and military capabilities to handle such a challenge, the Althea mission would face potential failure, and the competence of the EU's CSDP would come under increasing question. In such an eventuality, Washington may be pressed by its European allies to participate in deploying ground forces to restore stability. If violence escalates significantly, calls for peacemaking or peacekeeping missions will increase in the region and in several European capitals. The U.S. administration would then have to weigh the costs and benefits of recommitting ground forces in the Western Balkans.

Russia's Interventions.

Russia sees an opportunity to expand its reach in the "European space" given that the EU is beset by economic crisis and political indecision, with uncertain prospects for further enlargement beyond Croatia. Concurrently, NATO's further expansion in the Western Balkans, beyond the absorption of Montenegro, remains on hold. Macedonia is blocked, Serbia is opposed, Bosnia-Herzegovina is disunited, and Kosova is ineligible. Meanwhile, the United States is focused on other regions of the world, and its disengagement can weaken NATO's impact in Europe. As a result, Moscow seeks to intensify its political influence, particularly among states with no immediate prospect for Western integration, by employing three key tools: diplomatic assertiveness, conflict prolongation, and economic dependence.

Moscow is outspoken in support of Serbia, especially in its struggle over Kosova and in blocking Prishtina's membership in major international institutions such as the UN and the OSCE. Serbia remains the Kremlin's most reliable political link in the region, not because of any Slavic-Orthodox fraternity but as a consequence of cold political calculation. Belgrade has consistently appealed to Russian solidarity, whether over preserving Yugoslavia's integrity, creating a Greater Serbia, or retaining control over Kosova. Moscow in turn exploits Serbia's grievances against the United States and NATO to demonstrate that Russia remains a major factor in European affairs and in resolving intra-European disputes. Such symbiosis has proved beneficial for both capitals.

The Kremlin perceives Serbia as a useful proxy in the middle of the Balkans, and the country is promoted as a bastion against American influence. Moscow has increased its presence in Serbia during recent years, especially by exploiting its support for Belgrade's position on Kosova to keep Serbia outside NATO. In a display of strategic blackmail, Russia's ambassador to NATO threatened that Moscow would reconsider its attitude toward Kosova's statehood if Belgrade petitioned to join NATO.[216] In sum, Kosova remains a useful bargaining chip for the Kremlin in undermining European unity and maintaining close relations with Belgrade.

On November 2, 2011, Serbia's President Boris Tadic sharply criticized a speech by the Russian ambassador to Belgrade, Aleksander Konuzin, at a nationalist rally and accused him of meddling in Serbia's internal affairs. [217] Six weeks after he blasted Serbian politicians at a Belgrade security forum for leaving it to Russia to defend Serbia's interests abroad, Konuzin praised the nationalist Serbian Progressive Party (SNS) at a rally in Belgrade on October 29, 2011. The Russian Foreign Ministry expressed its full support for Konuzin, saying his participation at the rally was "normal diplomatic practice." The ambassador also threatened Serbia with a downgrading of relations if Belgrade made major energy and other economic deals with third countries. The Serbian media also revealed the existence of agents of influence working within Serbian institutions and businesses on behalf of Russian interests.[218]

The Kremlin would also prefer that Serbia remain outside the EU and avoid its onerous legal standards in business transparency that would affect the operations of opaque Russian companies. Instead, Moscow proposes that Serbia join its planned Eurasian economic bloc, a centerpiece of Putin's approach toward the former Soviet Union. The Serbian media reported Moscow's plans for EU expansion by 2020 to include countries such as Serbia, together with other states excluded from the EU.[219] The EU purportedly plans to have centers in St. Petersburg, Kyiv, Almaty, and Belgrade.

In an indication of the importance that Moscow assigns to Serbia, Putin visited Belgrade on March 23, 2011, to discuss boosting Russian investment through the planned South Stream natural gas pipeline. He reiterated Moscow's opposition to Kosova's independence and Serbia's NATO membership. In November 2011, more than 50,000 Kosova Serbs petitioned the Russian Embassy in Belgrade to grant them Russian citizenship.[220] According to Zlatibor Dzhordzhevich, leader of the Old Serbia movement that organized the petition, "dual citizenship would prevent the further Islamization of the Kosovo Serbs."[221] Volunteers traveled throughout Kosova listing names of those wishing to obtain Russian citizenship. In reality, Kosova Serbs have sought political support from Russia but have no intention of relocating to Russia. Nonetheless, Foreign Minister Sergei Lavrov jumped at the opportunity by claiming that Russia would examine requests for citizenship, describing the position of Serbs in Kosova as desperate.

In terms of conflict prolongation, the limited international recognition of Kosova has provided Russia with an opportunity to depict itself as the defender of international legality and the promoter of multilateralism, state sovereignty, and territorial integrity. Concurrently, it also promulgates the thesis of a pan-Albanian fundamentalist menace in attempts to forge pan-Slavic Orthodox unity under Russian patronage throughout the Western Balkans.

Moscow has also focused on the struggle over Bosnia-Herzegovina by supporting the leaders of the Serbian entity in their determination to resist streamlining the state and providing greater powers to the central government in Sarajevo. Having recognized the independence of two separatist regions in Georgia, Abkhazia, and South Ossetia, in August 2008, Russia retains the option of recognizing Bosnia's autonomous RS as an independent state. The Russian government is widely perceived to be supporting President Dodik and encouraging Banja Luka to undermine the authority of the OHR and to promote an American withdrawal.

Moscow employs two parallel tracks toward Bosnia-Herzegovina: an overt policy that recognizes Bosnia's state integrity and independence, and a covert policy that strengthens relations with the RS. The latter may enable Moscow to block Bosnia's entry into NATO. Moscow expands its influence by manipulating four tools: economic contracts, energy dependence, political corruption, and obstruction of Western initiatives. By exacerbating the prospect of RS secession, it intends to maintain Bosnia as a frozen or paralyzed state that can generate long-term problems for Washington and Brussels.

Through its vehement opposition to U.S. policy over Kosova and Bosnia-Herzegovina, Russia's government contributes to prolonging disputes and uncertainties within the region. The calculation is that Western preoccupation with inter-ethnic reconciliation and state building will dissipate and even terminate the region's integration into NATO and the EU. This will serve to justify Kremlin contentions that NATO cannot guarantee European security and a new continental security structure is needed in which Russia would play a major role. In sum, conflict provides Moscow with political leverage to advance its state ambitions.

The third Kremlin tool in the Balkans is the promotion of economic dependence by deploying energy resources, state loans, and business investments to gain political inroads. Plans to build major energy transportation systems between the Black Sea and the Adriatic Sea and Central Europe place the Balkans at the center of Russia's south European strategy. Moscow seeks to monopolize the supply of gas and oil passing through the region to Western Europe. Supply contracts and investment incentives provide significant inroads in a targeted country's economy and substantial influence over its foreign policy. The planned South Stream pipeline is calculated to place Serbia and Bulgaria at the center of Russia's ambitions and prevent the construction of a European energy network linking Central Asia, the South Caucasus, and Europe outside of Russia's control.

Russia's state company Gazprom owns the major share of Serbia's NIS oil company, and Belgrade is eager to host the southern hub of the planned South Stream pipeline through which Moscow seeks to eliminate the West's Nabucco gas pipeline project. A protocol on the Serbian section of South Stream was signed in October 2009 with Gazprom holding 51 percent and Serbian Gas holding the remaining 49 percent of shares in the project company. The pipeline is planned to cross from Serbia into Hungary while Russia entices the RS, Croatia, and Slovenia, with the prospect of including them in South Stream. Construction of South Stream is planned to start by the end of 2012 and finish in 2015, although the project has been riddled with doubts over routes, costs, and sources of gas.

On May 29, 2012, during his first foreign trip after the elections, President Tomislav Nikolic was warmly received in the Kremlin by President Putin.[222] Nikolic was seeking an $800 million loan, as Russia had previously promised a $1 billion dispersal but only delivered $200 million. However, the new Serbian government needs to carefully consider the conditions of any loans, as Russia's objective is to control Serbia's energy infrastructure and develop pipeline projects across the Balkans.

Macedonia's blocked path toward NATO and EU accession provides additional opportunities for Russia to expand its regional influence. Moscow can pose as the defender of Macedonian interests against alleged Albanian separatism and Islamic radicalism. Moves to extend Albanian political rights in Macedonia through the Ohrid Framework accord have been depicted as a subversive anti-Slavic and anti-Orthodox operation with the goal of creating a Greater Albania and dissolving Macedonia.[223]

In June 2012, Prime Minister Gruevski spent several days in Russia and announced that Macedonia was determined to deepen economic ties with Moscow by obtaining investment in the energy sector and scrapping visa requirements for Russian citizens visiting for less than a year to boost tourism.[224] Russia and Macedonia may also establish a free trade zone tying Skopje with the Customs Union of Russia, Belarus, and Kazakhstan.[225] On July 3, 2012, Skopje and Moscow signed a draft agreement that would provide Macedonia with an arm of the South Stream gas pipeline.[226] Whether any of these projects come to fruition, the promise of economic bounties lures cash-strapped Balkan governments into closer ties with Moscow, raises their susceptibility to political pressure, and undercuts the involvement of Western companies.

Two Russian state companies, Gazprom and Rosneft, have aimed to penetrate Albania's energy industry by purchasing state-owned energy companies. Moscow also wants to use Croatia to facilitate its energy penetration in the Adriatic. Gazprom has offered lucrative deals to Zagreb to export Russian oil through the Adria oil pipeline instead of Middle Eastern oil to Central Europe. Such a reversal would reduce Central European access to international oil markets, leaving them more dependent on Russian supplies. By offering an extension from South Stream to Croatia, Gazprom also intends to block the Adria Liquefied Natural Gas (LNG) terminal project on Krk Island to prevent it from undercutting Gazprom's monopolistic ambitions.

Moscow has criticized the Montenegrin government for its NATO membership bid and has threatened to curtail traditionally friendly relations between the two countries.[227] The Russian presence in Montenegro has grown in recent years and includes the purchase of real estate along the coast and in the capital Podgorica, Montenegro's large aluminum factory KAP, a bauxite mine in Niksic, and a string of gas stations.[228] Through these purchases, Russian oligarchs obtained greater influence over the country's economy. Russian immigration has also mushroomed, and several thousand Russian citizens are believed to reside in a country with a population of 620,000 inhabitants.[229]

Several Montenegrin parliamentarians have criticized the government for allowing Russian companies to purchase enterprises at low rates during the privatization process. By 2007, Russian investment comprised 27 percent of the foreign total. The global economic crisis after 2008 curtailed Russia's economic involvement and undermined trust in the commitment of Russian investors who failed to fulfill their contracts and pay workers' wages. The Montenegrin government needed to bail out several companies to prevent major unemployment and social turmoil, including the KAP aluminum plant, the biggest contributor to the country's GDP, which faced bankruptcy after its purchase by a Kremlin-connected oligarch.

Outside the energy sector, Russian companies have been particularly active in acquiring telecommunications, engineering, and banking interests in South East Europe.[230] For instance, in August 2012, Russia's largest bank, Sberbank, purchased the Austrian based financial conglomerate Volksbank International AG Group and was poised to take over Volksbank's branches in Bosnia-Herzegovina, Croatia, and Serbia.[231]

VII. WEST BALKAN CONFLICTS: CAUSES AND CONSEQUENCES

This chapter summarizes the policy challenges facing the Western allies in the Western Balkans. It outlines the factors that engender tensions, disputes, and unrest, and considers a number of existing and emerging conflict scenarios. In sum, 11 years after the last armed conflict in the former Yugoslavia was extinguished in the summer of 2001, long-term stability and security in the region is not fully assured and can be threatened by a number of internal and external factors.

The region can descend into a gray zone where splutters of progress in pursuit of reform are followed by prolonged periods of stagnation or even reversal. Such conditions provide fertile terrain for varieties of political radicalism and nationalist extremism. Although these are unlikely to generate outright war, they can create pockets of insecurity, conflict, and violence that would disqualify several states from EU membership. Such exclusion would in turn prolong and exacerbate local disputes and place enormous strain on international actors. It may even pull Washington back into the region through more intrusive diplomacy and mediation or even participation in new military missions.

The following challenges to state stability and regional security need to be carefully monitored and addressed.

International Exclusion.

There are two major alternatives for the Western Balkan states—Europeanization or marginalization. Europeanization means entry into both the EU and NATO, as was the case with Central Europe, the two East Balkan countries, and Slovenia and Croatia from the former Yugoslavia. If there are long delays in accession or the prospect disappears altogether, there will be little incentive for further reform to meet international standards. This could assign the region to a peripheral gray zone, increase opportunities for cross-border criminal organizations, result in depopulation as locals seek to escape to the EU, and radicalize a younger generation facing declining opportunities for employment and material advancement.

European Union Shortcomings.

While the EU's limitations as a hard power have been evident in its disjointed foreign policies and restricted military capabilities, its political and economic model may also be fading as an instrument of attraction. Skepticism about the future of the EU has grown within member states as the budgetary squeeze and debt crisis in several EU Mediterranean countries has generated profound economic uncertainty. Disappointment is also visible inside the EU over the performance of recent members, as well as with several older members encumbered by massive sovereign debts. Enlargement exhaustion among EU publics promotes reform fatigue among aspirant countries. Publics are becoming frustrated with seemingly endless entry requirements and prolonged timetables. Moreover, the long-term benefits of EU membership will also be seriously questioned if the Union's economic crisis deepens, and the EU itself appears to be splintering.

Economic Distress.

Economic problems are common to all Western Balkan states. These include the impact of the global credit crunch and economic recession, which has curtailed foreign investment, limited the availability of bank credits, and reduced worker remittances. Government debt, budget deficits, unemployment rates, and the number of citizens below the poverty level have soared, while economic growth rates have either decreased or remained static.[232] For instance, the gross national debt in Albania almost reached 60 percent of GDP in 2012 and 42 percent in Montenegro; unemployment was registered at 45 percent in Kosova, 32 percent in Macedonia, 27 percent in Bosnia-Herzegovina, and 20 percent in Serbia; while 32 percent of the Macedonian population is registered below the poverty line. A prolonged economic downturn, coupled with receding EU benefits, can stimulate populism and nationalism in several states.

Deficient State Building.

International supervision has suffered setbacks in Bosnia-Herzegovina as Serb leaders have campaigned to roll back the integration process. Herzegovina Croats are also reviving their pursuit of entity status and even prospective unification with Croatia. The EU is seen to be faltering in restitching multiethnic states or helping establish authoritative central government institutions in former war zones. Kosova's persistent territorial division, despite the presence of both

NATO and EU missions, reinforces this perception. In sum, a decrease in the U.S. role will expose the EU to potential failure.

Democracy Deficits.

The political status quo controlled by entrenched parties, coupled with growing economic inequalities, limited judicial reform, the pervasiveness of official corruption, and unfulfilled economic and occupational expectations, all have a negative social impact.[233] They heighten public alienation, demoralization, resentment, and anger; increase crime and lawlessness; provide ammunition to new protest movements; and encourage political extremism. Such movements can scapegoat ethnic and religious minorities or raise demands for revising borders or acquiring territories and fuel conflicts with neighboring states and international players.

Inadequate Leadership.

At a time of economic distress, government coalitions remain weak and indecisive, and they lack broad public legitimacy. Perceptions of widespread favoritism, nepotism, and corruption, and the persistence of cronyism and clientelism, have resulted in falling confidence in political leaders. In many cases, there is an absence of political unity on vital national questions, which in Bosnia-Herzegovina, Kosova, and Macedonia is coupled with incomplete state building. Government failings and slow international integration contribute to convincing compact minority populations that territorial partition or merger with a neighboring state is a viable option.

Nationalist Surpluses.

Various forms of nationalism continue to pose challenges to democratic consolidation and regional cooperation. For instance, in its 2012 report on democratic developments, Freedom House underscored that nationalism in much of the Western Balkans hampers regional reconciliation and obstructs relations with the EU.[234] Populist and nationalist elements benefit from economic stagnation and public disaffection and advocate xenophobia as a solution to numerous domestic challenges. Nationalists thrive where government institutions are not viewed as fully legitimate, especially where state building is incomplete.[235] Nationalism and populism are also enabled when internally undemocratic parties win elections, emplace loyal supporters in key institutions, and prolong their rule through institutional, financial, and informational manipulation.[236] Religious radicalism has not been an important factor in the region, as the Muslim population is largely moderate, secular, and divided by ethnicity, language, and doctrine, but a pauperized minority may become susceptible to ultra-conservative anti-Western influences.

Generational Challenges.

One cannot assume that the younger generation with no immediate experience of war will not resort to conflict. Ethnic separation in education, employment, residence, and marriage, even in a single state such as Bosnia-Herzegovina and Kosova, can foster ignorance, fear, victimhood, and marginalization. In the midst of economic stagnation, this can generate feelings of injustice and anger and leave people susceptible to nationalist appeals. A comprehensive survey conducted in all former Yugoslav republics during 2011 concluded that differences between generations were less pronounced than differences between ethnic groups in shaping attitudes.[237] Moreover, inter-ethnic distrust is commonplace among the post-1991 generation, and most people do not view the Western Balkans as a single cultural space.

Citizens may also be increasingly drawn toward the pan-European alienation felt by many young people with restricted opportunities in EU member states. Unfulfilled expectations and thwarted ambitions can lead to the rejection of existing political structures and gravitation toward extremist movements. Young people, especially the less educated, are more gullible to manipulation through a naïve belief in conspiracy theories promulgated by radical political groups to gain adherents.

Conflict Scenario.

In the midst of the destabilizing factors outlined previously, ethnic tensions can escalate and spark flash-points of conflict in the Western Balkans. The following potential conflict scenarios should be considered, as they would prove the most threatening to regional stability and could escalate toward armed conflicts.

In Bosnia-Herzegovina, the standoff between the Serbian entity and the government in Sarajevo comes to a head. The RS entity representatives withdraw from Bosnian government institutions and stage a referendum on separation and independence. Such moves provoke calls among Bosniaks to eliminate the

Serbian autonomous region and centralize the state. Croats capitalize on the unrest to declare an autonomous region in western Herzegovina. This triangular radicalization leads to violent incidents and a war footing in various parts of the country in anticipation of further conflict.

In northern Kosova, Serbian minority leaders stage a referendum on independence and unification with Serbia and appeal to Belgrade for protection as Albanian Kosovars are radicalized by Serbian separatism and prepare for a showdown. In southern Serbia, Albanians in the Presevo valley boycott state institutions, stage rallies, declare an autonomous region, and announce that they seek unification with Kosova. The escalating division of Bosnia-Herzegovina encourages Bosniak Muslims in Serbia's Sandzak region to push for autonomy. In reaction, the Serbian government cracks down on Albanian activism in the Presevo valley and Bosniak agitation in the Sandzak. It also escalates its nationalist agenda by openly supporting the partition of Bosnia-Herzegovina and Kosova.

In Montenegro, the pro- and anti-independence divisions take on a territorial and ethnic dimension, with Serbs in the north of the country establishing a National Council and claiming separate status. The Montenegrin government moves to regain control over the region, while Serbian activists appeal to Belgrade for direct assistance.

In Macedonia, Albanian frustration with government policy and escalating inter-ethnic incidents leads to a collapse of the ruling coalition. Albanian leaders declare an autonomous region along the Albanian and Kosova border and call for a division of the country. Belgrade offers assistance to Skopje against Albanian militants, while Albania is drawn into the conflict to protect its ethnic kindred. Macedonian politics radicalize and the government declares the cancellation of the Ohrid agreements and the defense of Macedonia's territorial integrity.

Pursuing Partition.

Some Western Balkan politicians and analysts contend that international actors should not block the option of state partition and territorial exchange if this is acceptable to both sides in a dispute. While U.S. and EU policymakers oppose any further state divisions, some observers believe that the disintegration of Yugoslavia has not been completed. For example, trying to preserve Bosnia-Herzegovina as a single state or maintaining Serbian majority municipalities inside Kosova may be costly, conflictive, and ultimately counterproductive.

If violence is to be avoided, such partition proposals are only realistic where they are acceptable to all parties in the dispute and are seen as part of an amicable territorial exchange. It may be feasible to exchange the four northern municipalities of Kosova containing a Serbian majority for three southern municipalities in Serbia with an Albanian majority or to exchange minority populations between the two regions. However, this would require several conditions, including a bilateral agreement between the two governments recognizing each other's statehood, a public referendum and agreement by the affected populations to join a new state, appropriate compensation for civilians displaced by the arrangement who do not wish to join another state, and international supervision of the entire process to guarantee that it is conducted fairly and peacefully.

Without these conditions, the secession of RS from Bosnia-Herzegovina will spark demands for the separation of Sandzak from Serbia, and the secession of northern Kosova will stimulate demands for the separation of the Presevo valley from Serbia. This is likely to encourage other secessionist movements, whose leaders will calculate that the most effective strategy for success is to provoke violence and government retaliation, capture international media attention, and thereby gain the political initiative. Such scenarios will also encourage governments to stage crackdowns to prevent separatism, while nationalist militants may arm themselves on the pretext of defending national integrity and ensuring state survival.

Armed Militancy.

The prospect for full-scale war between states or proxy insurgencies directly sponsored by Western Balkan governments appears remote at present. Nonetheless, low-level armed conflicts, whether generated by insurgent groups, vigilante militias, or armed civilians, are possible in parts of the region. Bosnia-Herzegovina, Kosova, and Macedonia remain candidates for armed clashes if minority grievances expand and inter-ethnic discords escalate. For example, Kosova will require a NATO presence for several years to deter and prevent armed conflict, especially in the northern part of the country where the Serbian minority does not accept the legitimacy of the Kosova state.

With limited international deterrents, weak governments, growing ethnic polarization, and rising nationalism, armed groups may become active with covert support from some governments or opposition parties in neighboring states. The KLA can be resuscitated, with links to Albanian militias in southern Serbia, southern Montenegro, and western Macedonia. Serbian radicals can also recreate militia groups to protect the Serbian minority in Kosova, reinforce the RS in Bosnia-Herzegovina, establish a Serbian autonomous region in Montenegro, and attack Albanian targets in the Presevo valley. Macedonian nationalist militias can also be mobilized to defend the country's territorial integrity.

Terrorist Threats.

Spreading lawlessness, militia mobilization, and organized criminality enable terrorist cells to infiltrate the region. They can exploit growing militancy among various Muslim populations to gain recruits for attacks on government targets or foreign interests. The terrorist attack on Israeli tourists in Bulgaria on July 18, 2012, focused attention on a region that some observers view as a potential hub of anti-Western terrorism. Although militant Islamist influence in the Western Balkans is a marginal phenomenon, the extreme acts of individuals can upset intercommunal relations and provide ammunition to radicals. A few acts of terrorism, as in Sarajevo in October 2011 or in Burgas in July 2012, can misrepresent the Balkans as a major recruiting ground for jihadists. Such misperceptions feed the ambitions of nationalist leaders who claim that Wahhabism and Salafism are growing among Muslim populations and that they need to actively protect endangered Christian interests.

Additional Conflicts.

An additional source of conflict has been visible between Roma communities, present in most Western Balkan countries, and extremist members of the majority population. There is a potential for inter-ethnic violence provoked by poverty, organized racist attacks, and the creation of Roma self-defense groups. Violent anti-Roma attacks seen in 14 Bulgarian towns in the fall of 2011 can be replicated in other parts of the region. The Roma population is growing in the Western Balkans, with an estimated half a million in Serbia and 165,000 in Macedonia. The Roma are often the most vulnerable minority and the most obvious targets of police violence, forced evictions, societal discrimination, and verbal and physical harassment.[238] As living standards stagnate during the economic recession, Roma communities are likely to be scapegoated by local gangs. Conflicts with the Roma population will radicalize both victims and perpetrators and contribute to broader social turmoil and inter-ethnic tensions.

Neighborhood Factors.

If Greece leaves the Eurozone and its living standards fall precipitously, this would send a negative signal to all EU candidates in the Western Balkans and accentuate anti-enlargement sentiments within the EU. A potential social explosion in Greece can have an impact on the stability of neighbors. In the most damaging scenario, expanding impoverishment and ejection from the Eurozone will precipitate the emergence of an authoritarian government. Under the pretext of restoring order and defending national dignity, a nationalist regime could target minorities and neighboring states, thus generating conflicts with both Macedonia and Albania.

Russian Penetration.

Declining EU and U.S. influence, diminishing or receding prospects for EU and NATO integration, and falling Western investment would encourage more intensive Russian political penetration through diplomatic offensives, energy contracts, and business purchases. Some Western Balkan governments or separatist movements may welcome Moscow's support against internal and external threats, while diluting or discarding their integrationist orientations toward the EU and NATO.

Interstate Disputes.

Neighborhood relations can become increasingly conflictive. The Albanian government may succumb to nationalist and irredentist sentiments, whether toward unification with Kosova or in defense of allegedly threatened Albanian interests in Macedonia and over compensation from Greece for the expelled and dispossessed Cham population. This will heighten tensions between Tirana and Belgrade, Skopje, and Athens. Interstate disputes will undermine regional cooperation, foster radicalism, and subvert security throughout the Western Balkans.

In a further negative development, rival Balkan alliances may be formed, with a Russo-centric bloc including Serbia, Macedonia, and the breakaway Serbian autonomous regions of Kosova and Bosnia-Herzegovina, confronting a Western-centric bloc that includes Albania, Kosova, and the Muslim Bosniaks.

Lacking sufficient protection by either Washington or Brussels, the beleaguered governments may welcome assistance from various Muslim states, including a less secular and more ambitious Turkey, while reducing their European and American connections. Greece could also veer toward Russia if the bankrupt country leaves the Eurozone and Moscow solicits for its own naval base in the Mediterranean by offering funds and investments to a cash-strapped Athens. This could heighten tensions with several neighbors, including Turkey.

EU members, particularly Bulgaria, Romania, and Croatia, would also suffer the negative consequences of neighborhood instability and find themselves pulled into the regional struggle due to their ethnic or political connections with conflicted states. Such conditions can precipitate a wider ripple of insecurity in the Balkan-Black Sea region, with an impact on numerous trans-regional and pan-European factors, from state stability to economic development and energy security. All of these causes and consequences would also play a role in determining the future role of NATO as a security provider, the effectiveness of the EU as a pan-European institution, and the stature and position of the United States throughout South East Europe.

VIII. POLICY RECOMMENDATIONS

The following prescriptions and recommendations are offered to both Washington and Brussels in dealing with the uncertain political climate in the Western Balkans. The policy focus should combine several approaches, including deterring and preventing new rounds of destabilizing conflict while encouraging each government to pursue policies that consolidate the country's statehood, its democratic institutions, and its public legitimacy, regardless of the timetable for inclusion in either NATO or the EU.

Role of U.S. Military.

Uphold a Credible NATO Presence.

NATO is viewed in the West Balkan region as the only credible international military force. Additionally, NATO membership remains a barometer for enhancing national security, pursuing military modernization, and ensuring an effective trans-Atlantic alliance. Nevertheless, the attractiveness and impact of the Alliance could significantly weaken if the U.S. disengages militarily from the broader European theater to a level that reduces effective deterrents in the event of a national or regional emergency. In such conditions, NATO's European pillar will need to carry the burden of guaranteeing durable security in unsettled zones such as the Western Balkans, with uncertain consequences.

Maintain U.S. Engagement.

The current U.S. ground force deployment in NATO's KFOR mission in Kosova should be maintained until the new state has developed an effective indigenous defense force and its entire territory is under Prishtina's political authority. The U.S. presence within the NATO operation significantly contributes to deterring local actors from provoking conflicts and reassures the local population that their security is protected.

In the absence of a NATO mission with a U.S. component and without a stronger Kosova defense structure, local clashes between Albanians and Serbs could escalate and the likelihood will grow that vulnerable communities will arm themselves for self-defense or support militias that provide them with protection. This can lead to intensified violence that may necessitate the return of a more sizable and robust international force to disarm local militias and restore a semblance of security. Similar developments could also unfold in Bosnia-Herzegovina if clashes between Bosniaks and Serbs were to erupt in several parts of the country and exceed the capacities of the EU Althea force in defusing tensions and stemming conflict.

Assess Need for Military Deployments.

Any serious deterioration in regional security would increase the pressure on Washington from its European allies to participate in deploying fresh ground forces in Kosova or in Bosnia-Herzegovina, especially since the use of air power for the purpose of restoring stability at the local level would be largely ineffective. In effect, if violence escalates significantly, the calls for peacemaking or peacekeeping missions will increase in the region and in several European capitals. The U.S. administration would then have to weigh the costs and benefits of committing ground forces back into the Western Balkans.

Given current U.S. security priorities and budgetary constraints, Washington will expect its core European allies, either within a NATO context or acting in a cooperative coalition, to assume primary responsibility in reducing tensions and restoring peace. Although the United States would, in all likelihood, provide strong diplomatic, political, and logistical support for a European-led operation, there would be little appetite in the country for direct military engagement in the Western Balkans.

Decisionmaking on possible deployments would also depend on the effectiveness of any projected European military mission and whether violence and attempts at state partition can be contained by NATO's European pillar. In addition, a temporary suppression of violence may not guarantee long-term stability, political progress, or economic development, without which the affected states would further distance themselves from EU and NATO membership and sow the seeds for future unrest and conflict. In such conditions, the nature and degree of American involvement would remain a factor in policymaking, as Washington will need to closely monitor the broader dangers to European security stemming from any persistent failure to contain instability and institutionally integrate the Western Balkans.

Prevent Conflict Escalation.

Several priorities need to be pursued by Washington to preclude a scenario whereby U.S. ground forces are requested by European partners to participate in renewed peacemaking or peacekeeping missions in the Western Balkans. These policy priorities must include the following: more comprehensive strategic intelligence gathering; the identification and monitoring of local and foreign political actors promoting instability; early warning signals that can pinpoint and defuse impending conflicts; a strong Allied diplomatic response to any deterioration of political conditions or inter-ethnic relations in each Balkan state and at the interstate level; and a firmer trans-Atlantic strategic commitment to bringing all countries in the region into both NATO and the EU.

Primary Regional Objectives.

Regional Political Security.

The onus must be on incumbent governments forging interstate agreements that acknowledge each other's sovereignty, independence, and statehood, and accept common borders with no ambitions or pretensions to neighbors' territories. Each capital must clearly and openly revoke any support for irredentism toward nearby states.

Institutional Development.

Political institutions in each state must continue to be strengthened through competitive elections, the rule of law, official accountability, the separation of powers, citizens' participation, extensive minority rights, and the combating of official corruption, nepotism, clientelism, and organized criminality. While the EU has become a less magnetic force, democratic consolidation under the rule of law is beneficial for each country regardless of its eventual institutional destination.

Common Market.

To increase attractiveness for investors, the small economies of the Western Balkans need to deepen their coordination by creating a genuine common market. A combined regional effort can focus on four domains: energy, transport, environment, and trade. Such an initiative would also give fresh impetus to preparations for EU membership. It would create a larger market of almost 20 million people, where the free movement of goods, services, capital, and labor could bring immediate benefits instead of simply waiting for EU accession.

Business Growth.

Greater emphasis must be placed on developing an interconnected class of entrepreneurs that will help moderate the impulses of local nationalism. The initiative can be promoted through business loans, cross-border commerce, and joint corporations in specific sectors between neighboring states.

Energy Security.

The development of a common regional approach toward fossil fuels and renewable energy will help attract investors, connect Balkan and EU energy networks, promote the diversity of supplies, and generate more sustainable economic growth. Such a strategy will also undercut Russia's attempts at energy monopolization throughout South East Europe.

NGO Networking.

NGOs perform three important tasks that can be expanded. First, they mobilize citizens in a plethora of activities that empower their lives and improve their living conditions. Second, NGOs pursue high standards of government transparency and effectiveness by holding politicians fully accountable to the electorate. Third, they develop parallel networks for interstate cooperation in various domains, from human rights and consumer protection to architectural conservation and environmental activism.

Serbia-Kosova Relations.

Integrating Kosova.

The Belgrade-Prishtina relationship remains one of the keys to stability in the region, and northern Kosova remains a constant flashpoint. Both Brussels and Washington must focus on restoring Kosova's territorial integrity, as a divided Kosova will remain a permanent source of dispute. The EU-sponsored talks can become a mechanism and a catalyst for integrating the northern municipalities into the country's institutions. They must focus on promoting the rule of law in the northern municipalities and holding legitimate local elections.

Demarcating Serbia-Kosova Border.

NATO's KFOR mission, working in tandem with the KSF and local police, should be mandated to demarcate the border with Serbia. This frontier is notorious for smuggling and trafficking. The Kosova police have already started to take over responsibility for border protection from KFOR along the frontier with Albania and Macedonia. Integrated border management is a condition that Kosova must meet to be included in the EU's visa liberalization. Such an initiative will enhance regional security, particularly if joint border demarcations are agreed to with Serbian authorities.

Kosova's Development.

Expanding International Recognitions.

This approach must concentrate on the five EU nonrecognizers (Greece, Spain, Romania, Slovakia, and Cyprus). Otherwise, these states could block some of the outcomes of the Serb-Kosovar dialogue by disabling the EU from acting in unison. A more concerted EU position will have a positive impact on Kosova's domestic reform process and Prishtina's steps toward international integration.

Transitioning from Supervision to Accession.

Although formal international supervision over Kosova was terminated in September 2012, the dependency relationship with the United States and the EU may continue if Prishtina does not achieve a clear path toward both EU and NATO membership. Dependence on outside actors undermines domestic responsibility for policy implementation and limits political transparency, as the government will feel more accountable to foreign powers than its own citizens. It can also butress charges that the EU does not perceive Kosova as a candidate for integration and thereby engender Euroskeptic inclinations.

Establishing a Contractual EU Relationship.

The EU needs to initiate a contractual relationship between Prishtina and Brussels with specific conditions, roadmaps, and focused assistance, as has been the case in all other West Balkan states. This can lead to preparations for an SAA on track toward candidate status and provide the EU with greater leverage in Kosova because of the prospect of membership.

Bosnia-Herzegovina's Functionality.

Limiting Entity Vetoes.

In building a more effective and functional state, a precise list should be enumerated where entity vetoes can be applied vis-à-vis the state government. For instance, vetoes can be relevant in the event of war, states of emergency, or other forms of national danger, and must not be exploited to obstruct the functioning of central or entity administrations.

Supporting the Civic Option.

Western powers must support the civic option in Bosnian politics. Bosnia-Herzegovina as a divided tri-ethnic state dependent on foreign largesse is not a durable proposition. The alternative to a civic state that can wean itself from foreign dependence is to be partitioned into two or three ethno-national units, a process that can unleash armed conflicts and instigate regional instability.

Clarifying EU Integration.

The EU accession process needs greater clarity, momentum, and commitment. The EU delegation offices in Sarajevo must involve citizens in the EU project and enable them to pressure their leaders to implement necessary reforms. This requires a more extensive outreach program with the Bosnian public and mass media to inform citizens about the Union and encourage them to canvass for Bosnia's EU entry.

Maintaining an International Role.

A formula needs to be devised whereby either the OHR or the EUSR retains the powers necessary to forestall the weakening of state institutions and the secession of RS, while invigorating the reform program necessary for EU accession. Simply removing the OHR without an effective EU replacement may be a recipe for state fracture.

Ensuring Military Deterrence.

Emplacement of a small NATO unit in the Brcko district, supervised by an American civilian official, would send a strong political message and deter any moves toward land seizures and armed conflicts at this vital choke point between the two Bosnian entities.

Promoting Regional Dialogue.

Bosnia-Herzegovina needs to be involved in a trilateral dialogue with Serbia and Croatia to supplement the bilateral dialogue between Zagreb and Belgrade. This could be modeled on the Visegrad initiative in Central Europe and include an extensive agenda of consultation and mutual support.

Macedonia's Progress.

Mediating Name Dispute.

The dispute between Athens and Skopje over Macedonia's name remains deadlocked. An interminable delay in resolving the dispute can raise nationalist fervor in both countries and destabilize the Macedonian state. International actors need to influence both sides involved in the controversy by tying an agreement to financial and other incentives and underscoring the negative consequences of an indefinite impasse.

Dampening Polarizing Nationalism.

Closer attention must be paid to the simmering tensions between Albanian and Slavic Macedonians, as they may have grave consequences for the survival of the state. The focus should be on preventing any rollback of the Ohrid Framework Agreement while developing new channels of political and social coexistence between Macedonian and Albanian communities.

Alternative Recommendation.

With several EU representatives urging full "local ownership" of the state-building process and a less intrusive international role, the prospect of state fracture and partition may need to be taken more seriously. Arguments can be made that if the majority of the population in a distinct region within an existing state favors separation and statehood and holds a referendum on such a decision, then the territory in question should be allowed to secede. Such a process could entail the merger of separated territories with existing states, as well as territorial or demographic exchanges between neighboring countries.

However, such proposals are only feasible where they are acceptable to all parties in the dispute. They would therefore require several specific conditions, including a bilateral agreement between the two governments recognizing each other's statehood and independence and agreeing to specific border adjustments; an internationally monitored public referendum and agreement by the majority of the affected populations to join a new state; appropriate compensation and housing for civilians displaced by the arrangement who do not wish to join another state; and international supervision of the process of territorial transfer or population exchange to guarantee that it is conducted fairly and peacefully. Without such conditions and with a diminished international deterrence, the pursuit of territorial partition, merger, and expansion could precipitate a cascade of violence throughout the Western Balkans. This would once again ensnare international organizations and necessitate new military missions, whether in combat operations or peacemaking and peacekeeping assignments.

POSTSCRIPT

The EU-brokered agreement between Serbia and Kosova on April 19, 2013, was declared a breakthrough in "normalizing" relations between the two states. An Association of Serbian Municipalities is to gain limited autonomy while recognizing the overall authority of the Kosova state. The government in Prishtina claimed that the plan will bring the northern municipalities under the control of central institutions. However, political opponents view the arrangement as the thin end of the wedge toward Serbian autonomy similar to the Bosnian model. Meanwhile, Serbian minority leaders asserted that the denial of self-determination will either lead to a Serbian exodus or to local unrest. The struggle over Kosova's statehood and stability will continue as the agreement is implemented.

ENDNOTES

1. *The Failed State Index, 2010,* Washington, DC: The Fund for Peace, 2011, available from *www.fundforpeace.org*.

2. A survey of tensions along contested or undemarcated border areas can be found in "Drawing Boundaries in the Western Balkans: A People's Perspective," *Saferworld,* October 2011, available from *www.saferworld.org.uk/downloads/pubdocs/Drawing%20boundaries%20in%20the%20Western%20Balkans.pdf*.

3. See Veton Surroi, "The Unfinished States in the Balkans and the EU: The Next Wave," in Morton Abramowitz *et al., The Western Balkans and the EU: The Hour of Europe, Chaillot Paper,* No. 126, Paris, France: European Union Institute for Security Studies, June 6, 2011, p. 112, available from *www.iss.europa.eu/publications/detail/article/the-western-balkans-and-the-eu-the-hour-of-europe/*.

4. For a discussion of precedents for self-determination, see Florian Bieber, "The Western Balkans After the ICJ Opinion," in Abramowitz *et al., The Western Balkans and the EU: The Hour of Europe,* pp.135-144, available from *www.iss.europa.eu/publications/detail/article/the-western-balkans-and-the-eu-the-hour-of-europe/*.

5. "President: Without EU Serbia Can Sink Into Nationalism," *Tanjug,* Belgrade, Serbia, December 8, 2011, *www.b92.net/eng/news/politics-article.php?yyyy=2011&mm=12&dd=08&nav_id=77685*.

6. Kris Brown and Bojana Barlovac, "Regional Leaders Shun Nikolic's Inauguration," *Balkan Insight,* June 11, 2012, available from *www.balkaninsight.com/en/article/regional-leaders-decline-invitation-to-nikolic-s-inauguration*.

7. See Bojana Barlovac, "Serbia's New Leaders Agree on Cabinet of 15," *Balkan Insight*, June 29, 2012, available from *www.balkaninsight.com/en/article/serbia-govt-to-have-15-ministries*.

8. "Serbian Parliament Speaker Calls For Closer Russia Ties," *Radio Free Europe, Radio Liberty*, May 9, 2007, available from *www.rferl.org/content/article/1076353.html*.

9. "Serbian Orthodox Church on the Brink of Schism," *Helsinki Bulletin*, No. 71, November 2010, Belgrade, Serbia, Helsinki Committee for Human Rights in Serbia.

10. Igor Jovanovic, "Agreement Reached on Kosovo's Regional Representation," *Southeast European Times*, February 25, 2012, available from *setimes.com/cocoon/setimes/xhtml/en_GB/features/setimes/features/2012/02/25/feature-01*.

11. "EU Candidacy Postponed: A Test for Democratic Party," *Helsinki Bulletin*, No. 85, December 2011, Helsinki Committee for Human Rights in Serbia, Belgrade, Serbia.

12. "President Wants 'Historic Kosovo Solution'," *B92*, Belgrade, Serbia, March 18, 2011, available from *www.b92.net/eng/news/politics-article.php?yyyy=2011&mm=03&dd=18&nav_id=73301*.

13. "Dačić in Fresh Kosovo Partition Statements," *B92*, Belgrade, Serbia, May 18, 2011, available from *www.b92.net/eng/news/politics-article.php?yyyy=2011&mm=05&dd=18&nav_id=74401*.

14. "Kosovo Partition: A Mission Impossible," *Helsinki Bulletin*, No. 82, October 2011, Helsinki Committee for Human Rights in Belgrade, Serbia.

15. Testimony of Phil Gordon, Assistant Secretary, Bureau of European and Eurasian Affairs, U.S. Department of State, Subcommittee on Europe and Eurasia, U.S. Congress, November 15, 2011.

16. Andrew Rettman, "Sweden Urges EU To Take Control of North Kosovo Problem," November 28, 2011, available from *euobserver.com/15/114422*.

17. "Tadic Says His Plan for Kosovo is Realistic," *Tanjug*, Sarajevo, Bosnia-Herzegovina, February 4, 2012, available from *www.b92.net/eng/news/politics-article.php?yyyy=2012&mm=02&dd=05&nav_id=78634.*

18. "Kosovo Partition Would be Best Solution," *Beta*, Prishtina, Kosovo, February 5, 2012, available from *www.b92.net/eng/news/politics-article.php?yyyy=2012&mm=02&dd=05&nav__id=78634.*

19. For an overview of flashpoints, see "Kosovo and Serbia: A Little Goodwill Could Go A Long Way," *Europe Report*, No. 215, Brussels, Belgium: International Crisis Group, February 2, 2012, available from *www.crisisgroup.org/en/regions/europe/balkans/kosovo/215-kosovo-and-serbia-a-little-goodwill-could-go-a-long-way.aspx.*

20. Fatmir Aliu, "NATO to Shut Illegal Crossings in Kosovo," *Balkan Insight*, April 7, 2012, available from *www.balkaninsight.com/en/article/kfor-to-close-the-illegal-crossings-in-northern-kosovo.*

21. "UN Envoy Warns Of Escalating Kosovo Tensions," *Radio Free Europe/Radio Liberty*, November 30, 2011.

22. Bojana Barlovac, "KFOR Fears Kosovo Violence from Either Side," *Balkan Insight*, November 26, 2011, available from *www.balkaninsight.com/en/article/kfor-statements-on-violence-worry-kosovo-serbs.*

23. "Dacic: New Demarcation is Needed," *B92*, October 15, 2011, Belgrade, Serbia.

24. Fatos Bytyci, "Kosovo Accuses Serbia of Police Abduction," *Reuters*, March 31, 2012, available from *www.reuters.com/article/2012/03/31/us-kosovo-serbia-idUSBRE82U0D820120331.*

25. Fatmir Aliu, "Kosovo Albanians Want More Police in North," *Balkan Insight*, April 10, 2012, available from *www.balkaninsight.com/en/article/kosovo-albanians-demand-extra-police-presence-in-the-north.*

26. Fatmir Aliu, "Kosovo President Seeks Consensus on Crisis in North," *Balkan Insight*, April 11, 2012, available from *www.balkaninsight.com/en/article/jahjaga-call-s-an-urgent-security-meeting-about-the-north.*

27. See "Terrorist Attack on Albanians for the Next Negotiations with Serbia," Lëvizja Vetevendosje Statement, Prishtina, Kosova, April 8, 2012.

28. See Fatmir Aliu, "Albanians and Serbs Brawl in Kosovo Enclave," *Balkan Insight*, April 4, 2012, available from *www.balkaninsight.com/en/article/albanians-and-serbs-massive-brawl-in-gracanica*; "Attack in Pristina Aimed at Influencing Elections," *Tanjug*, April 4, 2012, available from *www.tanjug.rs/news/38064/attack-in-pristina-aimed-at-influencing-elections.htm*; and Gordana Igric, "Serbia's Childish Game of Tit-for-Tat," *Balkan Insight*, March 29, 2012, available from *www.balkaninsight.com/en/article/serbia-s-childish-game-of-tit-for-tat.*

29. Igor Jovanovic, "Serbia's Withdrawal from Elections in Kosovo Sparks Confusion, Defiance," *Southeast European Times*, April 17, 2012, available from *www.setimes.com/cocoon/setimes/xhtml/en_GB/features/setimes/features/2012/04/17/feature-01.*

30. "NATO Grants KFOR Request for Reinforcements," *Tanjug*, April 24, 2012, available from *www.b92.net/eng/news/politics-article.php?yyyy=2012&mm=04&dd=24&nav_id=79930.*

31. Fatmir Aliu, "Border Clashes Mar Battle of Kosovo Commemoration," *Balkan Insight*, June 29, 2012, available from *www.balkaninsight.com/en/article/32-kosovo-policemen-injured-in-clashes-with-serbs.*

32. Fatmir Aliu, Bojana Barlovac, "Stoning of Serb Buses in Kosovo Roundly Condemned," *Balkan Insight*, June 29, 2012, available from *www.balkaninsight.com/en/article/attack-on-serb-children-a-shameful-act.*

33. *Balkan Watch*, Vol. 13, No. 22, December 19, 2011.

34. Nedim Dervisbegovic, "Bosnia's Escalating War Of Words," *Radio Free Europe*, October 19, 2011.

35. For example, see Dodik's speech at the fourth session of the RS National Assembly on April 13, 2011, quoted in Vlado Azinovic, Kurt Bassuener, and Bodo Weber, *Assessing the Potential for Renewed Ethnic Violence in Bosnia and Herzegovina,* Atlantic Initiative, Washington, DC: Democratization Policy Council, October 2011, p. 19.

36. Consult "Brcko Unsupervised," *Policy Briefing,* No.66, Sarajevo, Bosnia; Istanbul, Turkey; and Brussels, Belgium: International Crisis Group, December 8, 2011.

37. "Bosnia's Muslims and Croats Against End of Brcko Supervision," May 18, 2012, available from *www.finchannel.com/Main_News/Politics/109649_Bosnia%27s_Muslims_and_Croats_against_end_of_Brcko_supervision_/.*

38. For a summary of Belgrade's pressures, see Srdan Kusovac, "Montenegro—the Next Member of the Alliance," *Democracy and Security in Southeastern Europe,* Atlantic Initiative, Sarajevo, Bosnia, Vol. 1, Issue 1, 2010, pp. 27-32.

39. "Montenegro/Serbia Politics: Deteriorating Relations," *The Economist Intelligence Unit,* May 17, 2011, available from *viewswire.eiu.com.*

40. Serbian Government General Secretariat, Government of Serbia Strategies, "The Strategy for Preserving and Strengthening Relations between the Parent Country and the Diaspora and the Parent Country and the Serbs in the Region," January 21, 2011, Vol. 05, No. 019-169/2011-002, available from *www.gs.gov.rs/doc/strategije/strategija_dijaspora0389_cyr.zip.*

41. "Conclusions on Montenegro," excerpt from the Communication from the Commission to the European Parliament and the Council, "Enlargement Strategy and Main Challenges 2011-2012," available from *ec.europa.eu/enlargement/pdf/key_documents/2011/package/strategy_paper_2011_en.pdf.*

42. Milena Milosevic, "Montenegro Welcomes Nikolic's Line on Independence," *Balkan Insight,* May 31, 2012, available from *www.balkaninsight.com/en/article/montenegro-welcomes-nikolic-s-softened-rhetoric.*

43. "Montenegro/Serbia Politics: Mending Ties," *The Economist Intelligence Unit,* October 19, 2011, available from *viewswire.eiu.com.*

44. See the *Census of Population, Households and Dwellings in Montenegro 2011,* Podgorica, Montenegro, Statistical Office, December 7, 2011.

45. Erhan Türbedar, "Montenegro 2011 Census Details Announced," available from *www.trtdari.com/trtworld/en/newsDetail.aspx?haberkodu=324efe55-d02f-49d1-ad1a-5eaff3b30f70.*

46. Milena Milosevic, "Mass Protest in Podgorica Calls Govt to Account," *Balkan Insight,* March 19, 2012, available from *www.balkaninsight.com/en/article/mass-protest-held-in-podgorica.*

47. Milena Milosevic, "Montenegrin Protesters Call on PM to Quit," *Balkan Insight,* May 16, 2012, available from *www.balkaninsight.com/en/article/montenegrin-protesters-repeat-their-demand.*

48. Milena Milosevic, "Montenegro to Open Embassy In Pristina," *Balkan Insight,* June 1, 2012, available from *www.balkaninsight.com/en/article/montenegro-to-open-embassy-in-kosovo.*

49. Milena Milosevic, "Montenegro Puts Ambassador To Kosovo on Hold," *Balkan Insight,* June 8, 2012, available from *www.balkaninsight.com/en/article/vujanovic-puts-ambassador-to-kosovo-on-hold.*

50. See the Population Census of 2002, released by The Statistical Office of the Republic of Serbia, available from *webrzs.stat.gov.rs/axd/en/popis.htm.*

51. "Poll Shows Rising Support for Serbian Province's Independence," BBC Worldwide Monitoring, December 30, 2011. Original source: R. D., "Number of Separatists in Vojvodina Growing," *Danas* website, December 27, 2011.

52. "Vojvodina's Statute: First Major Step Towards Modernization and Decentralization," Helsinki Committee for Human Rights in Serbia, No. 51, December 2009, Belgrade, Serbia, available from *www.helsinki.org.rs/doc/HB-No51.pdf.*

53. "Vojvodina Should Have Its Own Police," *B92,* March 12, 2012, available from *www.b92.net/eng/news/politics-article.php?yyyy=2012&mm=03&dd=12&nav_id=79226.*

54. Bojan Pajtic, "Official: No Separatism in Vojvodina," *B92,* October 3, 2010, available from *www.b92.net/eng/news/politics-article.php?yyyy=2010&mm=10&dd=03&nav_id=70050.*

55. "An Autonomous Vojvodina: Exit Strategy," *The Economist,* December 30, 2009, available from *www.economist.com/node/15180922.*

56. "Some Vojvodina Towns Refusing to Add Province's Flag to Serbian One," *BBC Worldwide Monitoring,* May 14, 2011; and "Refusing Dual Flags," *Vecernje Novosti* website, May 11, 2011.

57. Igor Jovanovic, "Controversy Erupts over Vojvodina's Brussels Office," *Southeast European Times,* August 22, 2011, available from *www.setimes.com/cocoon/setimes/xhtml/en_GB/features/setimes/features/2011/08/22/feature-04.*

58. Lily Lynch, "Serbian Court Disputes Limits of Vojvodina Autonomy," *Southeast European Times,* July 18, 2012.

59. "Bosnia 20 Years On: Remembrance but not Reconciliation," *TransConflict,* May 10, 2012, available from *www.insightonconflict.org/2012/05/bosnia-twenty-years-on.*

60. For a recent report, see Janusz Bugajski and Heather Conley, *A New Transatlantic Approach for the Western Balkans: Time for a Change in Serbia, Kosova, and Bosnia-Hercegovina,* A Report of the CSIS Europe Program, Washington DC: Center for Strategic and International Studies, November 2011.

61. A summary of Bosnia's predicament can be found in Marko Attila Hoare, "Bosnia and Herzegovina: The Crumbling Balkan Keystone," *Democracy and Security in Southeastern Europe,* Atlantic Initiative, Sarajevo, Bosnia, Vol.1, Issue 1, 2010, pp. 50-55.

62. Agency for Statistics of Bosnia Herzegovina, Demography 2010, October 2011, available from *www.bhas.ba/tematskibilteni/DEM_2010_001_01-bh.pdf*; UNDP Bosnia and Herzegovina, Hu-

man Development Report 2002, available from *hdr.undp.org/en/reports/national/europethecis/bosniaherzegovina/Bosnia_and_Herzegovina_2002_en.pdf.*

63. For sensationalist depictions of Bosnian Muslims and their political leaders, see John R. Schindler, *Unholy Terror: Bosnia and al-Qa'ida and the Rise of Global Jihad*, Grand Rapids, MI: Zenith Press, 2007; and Christopher Deliso, *The Coming Balkan Caliphate: The Threat of Radical Islam to Europe and the West,* Westport, CT: Praeger, 2007.

64. A valuable analysis of Wahhabism, Salafism, and the Takfiri movement in Bosnia can be found in Juan Carloz Antunez, "Comprehensive Analysis of the Phenomenon of Wahhabism in Bosnia-Herzegovina," London, UK: Bosnian Institute, September 2008, available from *www.bosnia.org.uk/news/news_body.cfm?newsid=2468.*

65. Eldar Sarajlic, "The Return of the Consuls: Islamic Networks and Foreign Policy Perspectives in Bosnia and Herzegovina," paper presented at the conference "After the Wahhabi Mirage: Islam, Politics, and International Networks in the Balkans," University of Oxford, European Studies Centre, December 2010, p. 9.

66. "Radical Groups in the Balkans: The Case of Wahhabi Jašarevic," *Helsinki Bulletin*, Helsinki Committee for Human Rights in Serbia, No. 84, Belgrade, Serbia, November 2011, available from *www.helsinki.org.rs.*

67. For an overview of Turkey's role in the Balkans, see Janusz Bugajski, "Turkey's Impact in the Western Balkans," Washington, DC: Atlantic Council, February 2012, available from *www.acus.org/files/TAR/030112_ACUS_Bugajski.PDF*; and Zarko Petrovic and Dusan Reljic, "Turkish Interests and Involvement in the Western Balkans: A Score-Card," *Insight Turkey,* Vol. 13, No. 3, 2011, pp. 159-172.

68. See "Destiny of Bosnia and Herzegovina: Twenty Years Later in Karadjordjevo," Ankara, Turkey: International Strategic Research Organization, November 2011, available from *www.usak.org.tr/EN/myazdir.asp?id=2104.*

69. "Turkey in the Balkans: The Good Old Days? Talk of an Ottoman Revival in the Region Seems Exaggerated," *The Economist,* November 5, 2011.

70. Based on the author's discussions with President Milorad Dodik and other RS officials in Banja Luka, Bosnia-Herzegovina, in June 2011.

71. Nina Brankovic, "Bosnia's Citizens: Getting More Cooperative in Administrative Matters, but Still Strongly Opposed in Soccer," *Balkananalysis.com,* January 30, 2011, available from *www.balkanalysis.com/bosnia/2011/01/30/bosnia%E2%80%99s-citizens-getting-more-cooperative-in-administrative-matters-but-still-strongly-opposed-in-soccer.*

72. See "Bosnia: What Does Republika Srpska Want?" *Europe Report,* No. 214, October 6, 2011, Brussels, Belgium: International Crisis Group.

73. Based on the views of several political leaders and analysts in Sarajevo during the author's visit in June 2011. According to opinion polls conducted in 2010, 81 percent of Serbs in the RS support the entity joining Serbia. See International Crisis Group, October 6, 2011, p.11.

74. Gallup polls indicate that 87 percent of Serbs in Bosnia favor secession and 43 percent of Croats support splitting the Bosnian Federation into two entities. Denisa Sarajlic-Maglic, "BiH After the Elections - A Tale of Disillusioned Optimism," in Abramowitz *et al., The Western Balkans and the EU: The Hour of Europe,* p. 54, available from *www.iss.europa.eu/publications/detail/article/the-western-balkans-and-the-eu-the-hour-of-europe/.*

75. "The Battle for Mostar," *SM Global Politics and Business Intelligence Review,* Vol. 1, No. 1, May 2012,Washington, DC.

76. Elvira Jukic, "Mostar's Unity Not in Danger, Inzko Says," *Balkan Insight,* March 19, 2012, available from *www.balkaninsight.com/en/article/international-community-mostar-cannot-be-disintegrated.*

77. Drazen Remikovic, "Islamic Community in BiH to Appeal RS Court Verdict on Mosques," *Southeast European Times*, February, 16, 2012, available from *www.setimes.com/cocoon/setimes/xhtml/en_GB/features/setimes/features/2012/02/16/feature-02*.

78. Anes Alic, "International Community Pushes for Entity Demarcation in BiH," *South East European Times,* May 5, 2012.

79. "Hooligans Riot in Bosnian Town after Croatia's EURO Defeat," *B92*, June 19, 2012, available from *www.b92.net/eng/news/region-article.php?yyyy=2012&mm=06&dd=19&nav_id=80841*.

80. Julie Brethfeld and Nino Vadakaria, "Leaving the Past Behind. The Perceptions of Youth in Bosnia and Herzegovina," Sarajevo, Bosnia: Nansen Dialogue Center and Saferworld, March 2012, available from *www.saferworld.org.uk/smartweb/resources/view-resource/633*.

81. Based on private meetings with analysts and officials in Sarajevo in June 2011.

82. Azinovic, Bassuener, and Weber.

83. Consult Juan Carlos Antunez, "Wahhabism in Bosnia-Herzegovina and its Links to International Terrorism," Islam in South East Europe Forum, 2008, available from *iseef.net*; and Eldar Sarajlic, p. 20.

84. Elvira Jukic, "Islamist Death Threats Force Out Bosnia Minister," *Balkan Insight,* February 14, 2012, available from *www.balkaninsight.com/en/article/dignity-rather-than-chair-sarajevo-slogans-say*.

85. Dusan Babic, "Separating Religion and State in Bosnia," *TransConflict,* November 17, 2011, available from *www.transconflict.com/2011/11/separating-religion-and-state-in-bosnia-711*.

86. See Ulrich Heider, "A Country of Deadly Potential," *Democracy and Security in Southeastern Europe,* Atlantic Initiative, Vol. 1, Issue 1, Sarajevo, Bosnia, 2010, pp. 65-68.

87. Azinovic, Bassuener, Weber, pp. 36-38, 53-56.

88. Background material on the Sandzak can be found in Kenneth Morrison, "Political and Religious Conflict in the Sandzak," Balkans Series, 08/13, Wiltshire, UK: Advanced Research and Assessment Group, Defense Academy of the United Kingdom, April 2008.

89. For population statistics, see *en.wikipedia.org/wiki/Sandžak.*

90. Mico Tatalovic, "Fight Erupts Over New Bosniak Academy of Sciences," *Science Insider,* June 16, 2011, available from *news.sciencemag.org/scienceinsider/2011/06/fight-erupts-over-new-bosniak.html.*

91. See Justin Vela, "Serbia's Sandzak at Odds with Belgrade," *Transconflict,* December 7, 2010, available from *www.transconflict.com/2010/12/serbia-sandzak-at-odds-with-belgrade-712/.*

92. Muhamed Jusić, "Muslims in Serbia: Two Parallel Islamic Communities in Serbia. The Danger of Politicization of Religion," December 5, 2008, available from *www.globaliamagazine.com/?id=509.*

93. See Stephen Schwartz, "Islamist Terrorism in Bosnia as Turkish Interference Continues in the Balkans," *The Weekly Standard,* November 3, 2011, available from *www.weeklystandard.com/blogs/islamist-terrorism-bosnia-turkish-interference-continues-balkans_607715.html.*

94. "Sandzak: Still a Vulnerable Region," Helsinki Committee For Human Rights in Serbia, *Annual Report 2004,* Belgrade, Serbia.

95. See Juan Carloz Antunez, "Comprehensive Analysis of the Phenomenon of Wahhabism in Bosnia-Herzegovina," London, UK: Bosnian Institute, September 2008, available from *www.bosnia.org.uk/news/news_body.cfm?newsid=2468.*

96. "Radical Groups in the Balkans: The Case of Wahhabi Jašarevic," *Helsinki Bulletin,* Helsinki Committee for Human Rights in Serbia, No. 84, Belgrade, Serbia, November 2011, available from *www.helsinki.org.rs.*

97. Associated Free Press (AFP), October 29, 2011.

98. For background on Sandzak politics, see Beata Huszka, "Decentralisation of Serbia: The Minority Dimension," *CEPS Policy Brief*, No. 137, July 2007, Brussels, Belgium: Centre For European Policy Studies.

99. Erhan Türbedar, "Montenegro 2011 Census Details Announced," available from *www.trtdari.com/trtworld/en/newsDetail.aspx?haberkodu=324efe55-d02f-49d1-ad1a-5eaff3b30f70*.

100. For recent population estimates, see the Central Intelligence Agency, *The World Factbook*, available from *https://www.cia.gov/library/publications/the-world-factbook/*.

101. For material on the Greater Albania question, see Mimoza Ardolic, "Greater Albania—The Next Crisis in the Balkans?" Vaxjo, Sweden: Vaxjo University, Spring 2009.

102. "Conclusions on Albania," Excerpt from the Communication from the Commission to the European Parliament and the Council, Enlargement Strategy and Main Challenges 2011.

103. "Heading Home Again: Worried Albanians in Northern Greece Prepare to go Home," *The Economist*, January 14, 2012, available from *www.economist.com/node/21542818*.

104. Benet Koleka and Renee Maltezou, "For Albanians, Taste of Capitalism Turns Sour in Greece," *Reuters*, April 6, 2012, available from *www.reuters.com/article/2012/04/06/albania-greece-migrants-idUSL6E8F38XL20120406*.

105. Besian Bocka, "New Force Vows to Break Mould in Albania," *Balkan Insight*, January 12, 2012.

106. An assessment of the role of religion in Albanian politics can be found in Gezim Krasniqi, "The 'Forbidden Fruit': Islam and Politics of Identity in Kosovo and Macedonia," *Southeast European and Black Sea Studies*, Vol. 11, No. 2, June 2011, pp. 191-207.

107. Gjergj Erebara, "Imam's Dismissal Divides Muslims in Albania," *Balkan Insight*, March 14, 2012, available from *www.balkaninsight.com/en/article/imam-s-dismissal-sparks-row-in-albania-s-muslim-community.*

108. Available from *lajme.shqiperia.com/lajme/artikull/iden/1047176626/titulli/Kuvendi-i-Kosoves-me-rezolute-per-fundin-e-mbikeqyrjes.*

109. Based on conversations with EU officials in Washington, DC, July 2012.

110. "A Comprehensive Vision for the North: The Final Countdown," Policy Paper, Prishtina, Kosovo, February 2012.

111. Igor Jovanovic and Muhamet Brajshori, "Serbia Disputes ICG Claim: We Have no Security Forces in Kosovo," *Southeast European Times,* January 27, 2012, Belgrade, Serbia, and Prishtina, Kosovo.

112. "Referendum in North: 99.74% say 'No' to Priština," *B92*, February 16, 2012, available from *www.b92.net/eng/news/politics-article.php?yyyy=2012&mm=02&dd=16&nav_id=78814.*

113. "Conclusions on Kosovo," Excerpt from the Communication from the Commission to the European Parliament and the Council, *Enlargement Strategy and Main Challenges 2011-2012.*

114. Bojana Barlovac, "Kosovo, Serbia Reach Deal on Regional Representation," *Balkan Insight*, February 24, 2012, available from *www.balkaninsight.com/en/article/kosovo-serbia-reach-deal-on-regional-representation.*

115. Muhamet Brajshori and Bojana Milovanovic, "Misunderstandings Mire Kosovo Representation Agreement," *Southeast European Times,* Prishtina, Kosovo, and Belgrade, Serbia, March 16, 2012.

116. Based on a series of discussions with activists and political leaders in Prishtina in the spring of 2012.

117. Based on private discussions with the author in Prishtina in March 2012.

118. Anita McKinna, "The Vetëvendosje Movement in Kosovo: An Increasing Focus on Nationalism," *Balkananalysis.com,* February 22, 2012, available from *www.balkanalysis.com/kosovo/2012/02/22/the-vetevendosje-movement-in-kosovo-an-increasing-focus-on-nationalism.*

119. See Gezim Krasniqi, "The 'Forbidden Fruit': Islam and Politics of Identity in Kosovo and Macedonia," paper presented at the conference "After the Wahhabi Mirage: Islam, Politics, and International Networks in the Balkans," European Studies Centre, University of Oxford, December 2010.

120. "Growing Religiosity Sparks Kosovo Debate," *On Islam & Newspapers,* April 12, 2012, available from *www.onislam.net/english/news/europe/456605-growing-religiosity-sparks-kosovo-debate.html.*

121. Vehbi Kajtazi and Bekim Dobra, "Masked Individuals Threaten Public Order in Kosovo," *Koha Ditore,* Prishtina, Kosova, April 14, 2011.

122. See Linda Karadaku, "Extremist Organizations Still Spark Fear in the Balkans," *Southeast European Times,* March 23, 2012, available from *www.setimes.com/cocoon/setimes/xhtml/en_GB/features/setimes/features/2012/03/23/feature-01.*

123. Fatmir Aliu, "Kosovo Probes Claims of Armed Albanians Patrolling North," *Balkan Insight,* April 20, 2012, available from *www.balkaninsight.com/en/article/prosecution-on-hold-for-the-armed-kosovo-albanians-in-the-north.*

124. "Kosovo: Serb Returnees Threatened with Expulsion," *B92,* May 15, 2012, available from *www.b92.net/eng/news/politics-article.php?yyyy=2012&mm=05&dd=15&nav_id=80259.*

125. "Tensions Serious Risk for Kosovo's Stability," *Kosovo Compromise,* May 15, 2012, available from *www.kosovocompromise.com/cms/item/latestnews/en.html?view=story&id=3662&sectionId=1.*

126. Fatmir Aliu, "Serb Nationalists Threaten Mitrovica Albanians," *Balkan Insight*, April 17, 2012, available from *www.balkaninsight.com/en/article/obraz-gets-to-mitrovica*.

127. For an analysis of the Presevo valley conflict, see "Serbia: Maintaining Peace in the Presevo Valley," *Europe Report*, No. 186, Brussels, Belgium: International Crisis Group, October 16, 2007.

128. "Russia Opens 'Humanitarian' Base in Serbia," *Euractiv*, October 18, 2011, available from *www.euractiv.com/enlargement/russia-opens-humanitarian-base-s-news-508382*.

129. "Albanians Held in Serbia Over 'War Crimes'," *RIA Novosti*, Sarajevo, Bosnia, May 4, 2012.

130. Jovana Gec, "Serbia Police Arrest Eight in Tense South," *Associated Press*, May 4, 2012.

131. Safet Kabashaj, "Albanian Guerrilla Leaders Issue Warnings," *South East European Times*, May 18, 2012, available from *www.setimes.com/cocoon/setimes/xhtml/en_GB/features/setimes/features/2012/05/18/feature-03*.

132. Sinisa Jakov Marusic, "Macedonia Still Lags on Employing Minorities," *Balkan Insight*, April 4, 2012, available from *www.balkaninsight.com/en/article/macedonia-lags-with-minority-workers-employment*.

133. Sinisa Jakov Marusic, "Macedonia Disappoints Albanians on Flag Celebration," *Balkan Insight*, March 28, 2012, available from *www.balkaninsight.com/en/article/macedonia-won-t-celebrate-albanian-flag-day*.

134. Muhamet Brajshori, "Kosovo Looks to Keep Radical Elements at Bay," *Southeast European Times*, Prishtina, May 28, 2012, available from *www.setimes.com/cocoon/setimes/xhtml/en_GB/features/setimes/features/2012/05/28/feature-04*.

135. "Muslim Leaders in Macedonia Urge their Community to Tone Down Protests over Local Carnival," *Associated Press*, Skopje, Macedonia, January 30, 2012.

136. Sinisa Jakov Marusic, "Macedonia to Curb Fans' Offensive Chants," *Balkan Insight*, February 17, 2012, available from *www.balkaninsight.com/en/article/macedonia-curbs-fan-s-offensive-chanting*.

137. Safet Kabashaj, "Albanian Guerrilla Leaders Issue Warnings," *South East European Times*, May 18, 2012, available from *www.setimes.com/cocoon/setimes/xhtml/en_GB/features/setimes/features/2012/05/18/feature-03*.

138. "Macedonia: Ten Years After The Conflict," *Europe Report*, No. 212, Brussels, Belgium: International Crisis Group, August 11, 2011, p. 20.

139. Saso Ordanoski, "The Story of Macedonian Populism: 'All We Want is Everything'," in Abramowitz *et al.*, p. 100, available from *www.iss.europa.eu/publications/detail/article/the-western-balkans-and-the-eu-the-hour-of-europe/*.

140. Erhan Türbedar, "Montenegro 2011 Census Details Announced," available from *www.trtdari.com/trtworld/en/newsDetail.aspx?haberkodu=324efe55-d02f-49d1-ad1a-5eaff3b30f70*.

141. For an informative report, see Cafo Boga and Stefan Wolff, "Albanians in Montenegro: Waiting for Godot?" *Illyria*, No. 2067, July 12-14, 2011, available from *www.stefanwolff.com/files/Albanians%20in%20Montenegro.pdf*.

142. Background on Chameria can be found in Miranda Vickers, "The Cham Issue: Albanian National and Property Claims in Greece," April 2002, Sandhurst, UK: Conflict Studies Research Centre, available from *www.papers.ssrn.com*.

143. Based on the author's conversations with Cham representatives in Tirana in June 2012.

144. Based on the author's discussions in Tirana with leaders of the Red and Black Alliance in February 2012.

145. Klaudija Lutovska, "Minorities Say Albania Discriminates Against the Right to Nationality," *Southeast European Times*,

December 30, 2011, available from *www.setimes.com/cocoon/setimes/xhtml/en_GB/features/setimes/features/2011/12/30/feature-03.*

146. Areti Kotseli, "4,000 'Cham' Albanians Hold Protest in the Greek-Albanian Borders Claiming 'their' Land Back," *Greek Reporter*, July 1, 2012, available from *greece.greekreporter.com/2012/07/01/4000-cham-albanians-hold-protest-in-the-greek-albanian-borders-claiming-their-land-back/.*

147. Based on the author's discussions with Cham leaders in Tirana in May 2012.

148. Available from *www.defencegreece.com/index.php/2012/03/leader-of-greek-minority-party-in-albania-vangelis-doules-there-is-no-tsamouria/.*

149. "Protests at Cuts on Eve of Bailout Decision," *Athens News*, Athens, Greece, February 20, 2012, available from *www.athensnews.gr/portal/1/53463.*

150. For a recent analysis of Macedonia's political problems, see Janusz Bugajski, "The Republic of Macedonia/Former Yugoslav Republic of Macedonia," *Western Balkans Policy Review, 2010*, Washington DC: Center for Strategic and International Studies (CSIS), 2011, pp. 82-89.

151. See Saso Ordanoski, "The Story of Macedonian Populism: 'All We Want is Everything'," in Abramowitz *et al.*, pp. 95-109, available from *www.iss.europa.eu/publications/detail/article/the-western-balkans-and-the-eu-the-hour-of-europe/.*

152. Sinisa Jakov Marusic, "Macedonia Street Renaming Denounced as 'Revisionist'," *Balkan Insight*, June 18, 2012, available from *www.balkaninsight.com/en/article/renaming-of-skopje-streets-causes-jitters.*

153. See Denko Maleski, "Law, Politics, and History in International Relations: Macedonia and Greece," *New Balkan Politics*, Issue 12, 2010, available from *www.newbalkanpolitics.org.mk/documents/pdf/NBP,%20MaleskiD.pdf.*

154. Based on discussion with analysts in Skopje, Macedonia, in March 2012.

155. "Troika Presses Greece for More Cuts," *The Press Association*, July 27, 2012.

156. Based on discussion with officials and analysts in Skopje, Macedonia in March 2012.

157. Sinisa Jakov Marusic, "Greece 'FYROM-izes' Macedonia's Car Plates," *Balkan Insight*, June 21, 2012, available from *www.balkaninsight.com/en/article/greece-fyrom-izes-macedonian-car-plates*.

158. Sinisa Jakov Marusic, "Greece Says Macedonia Abused UN Chief's Visit," July 27, 2012, available from *www.balkaninsight.com/en/article/greece-macedonian-pm-abused-ki-moon-s-visit*.

159. For a valuable analysis, see "Macedonia: Ten Years After The Conflict," *Europe Report*, No. 212, Brussels, Belgium, International Crisis Group, August 11, 2011.

160. See Mirjana Maleska, "Inter-ethnic Relations in Macedonia: People Centred Analyses," *New Balkan Politics*, Issue 12, 2010, available from *www.newbalkanpolitics.org.mk/*.

161. Based on discussion with officials and analysts in Skopje, Macedonia, in March 2012.

162. "Conclusions on The Former Yugoslav Republic of Macedonia," Excerpt from the Communication from the Commission to the European Parliament and the Council, *Enlargement Strategy and Main Challenges 2011-2012*.

163. Sinisa Jakov Marusic, "EU, Macedonia Launch New Dialogue," *Balkan Insight*, March 16, 2012, available from *www.balkaninsight.com/en/article/eu-macedonia-launch-new-dialogue*.

164. "Macedonia Key Developments," *The Economist Intelligence Unit*, Country Report, November 1, 2011, available from *viewswire.eiu.com*.

165. Sinisa Jakov Marusic, "Envoy Alarmed by Macedonia Tensions," *Balkan Insight*, February 10, 2012.

166. "Muslim Leaders in Macedonia Urge their Community to Tone Down Protests over Local Carnival," Associated Press, Skopje, Macedonia, January 30, 2012.

167. Boris Babic, "Church Burns in Macedonia as Ethnic Tensions Flare Again," January 31, 2012, *DPA*, Berlin, Germany.

168. "Macedonian Embassy in Kosovo Suffers Molotov Cocktail Attack," *Novinite*, March 17, 2012, available from *www.novinite.com/view_news.php?id=137641*.

169. "Macedonian Police Establish Profile of Killers," *Balkan Insight*, April 15, 2012, available from *www.balkaninsight.com/en/article/macedonian-police-establish-profile-of-killers*.

170. Sinisa Jakov Marusic, "Macedonian Albanians Protest Police Arrests," *Balkan Insight*, May 4, 2012, available from *www.balkaninsight.com/en/article/albanians-protest-in-skopje-over-police-arrests*.

171. Misko Taleski, "Radical Islam Poses Regional Threat," *Southeast European Times*, May 10, 2012, available from *www.setimes.com/cocoon/setimes/xhtml/en_GB/features/setimes/features/2012/05/10/feature-03*.

172. Chris Deliso, "After Macedonia's Islamist Protest, Investigators Search for Significance Amidst Confusing Array of Motives and Clues," February 13, 2012, *Balkananalysis.com*, available from *www.balkanalysis.com/macedonia/2012/02/13/after-macedonias-islamist-protest-investigators-search-for-significance-amidst-confusing-array-of-motives-and-clues/*.

173. "Bulgarian Government Reacts Strongly to Macedonia's Hate Speech," August 9, 2012, *Novinite.com*, available from *www.novinite.com/view_news.php?id=142174*.

174. Sase Dimovski and Sinisa Jakov Marusic, "Macedonian Albanians Protest Over Police Killing," *Balkan Insight*, March 1, 2012, available from *www.balkaninsight.com/en/article/albanians-protest-after-macedonian-policeman-kills-two*.

175. Sinisa Jakov Marusic, "Ethnic Brawls on Buses Concern Maçedonia," *Balkan Insight*, March 9, 2012, available from *www.balkaninsight.com/en/article/skopje-bus-brawls-raise-ethnic-alarm*.

176. Linda Karadaku, "Extremist Organisations Still Spark Fear in the Balkans," *Southeast European Times*, March 23, 2012, available from *www.setimes.com/cocoon/setimes/xhtml/en_GB/features/setimes/features/2012/03/23/feature-01*.

177. Misko Taleski, "Macedonia, Albania Aim to Turn Page on Encyclopedia Dispute," *Southeast European Times* April 30, 2012, available from *www.setimes.com/cocoon/setimes/xhtml/en_GB/features/setimes/features/2012/04/30/feature-03*.

178. For an assessment, see *en.wikipedia.org/wiki/European_sovereign_debt_crisis*.

179. Information on the Summit is available from *www.g20.org/en*.

180. Niall Ferguson, "Why The Young Should Welcome Austerity," *BBC World News*, June 16, 2012, available from *www.bbc.co.uk/news/world-18456131*.

181. "Global Employment Trends 2012: Preventing a deeper jobs crisis," available from *www.ilo.org/global/research/global-reports/global-employment-trends/WCMS_171571/lang--en/index.htm*.

182. A recent economic overview can be found in "The Eurozone Crisis Affects the Transition Region," July 27, 2012, *Balkans.Com Business News*, available from *www.balkans.com/open-news.php?uniquenumber=152198&utm_source=MailingList&utm_medium=email&utm_content=jbugajsk%40csis.org&utm_campaign=Copy+of+22.07-.29.07.2012.+Wee+ENG+bd*.

183. François Heisbourg, "The Defense of Europe: Towards a new Transatlantic Division of Responsibilities," in Tomas Valasek, ed., *All Alone? What U.S. Retrenchment Means for Europe and NATO*, February 2012, London, p. 35.

184. *Ibid*., p. 38.

185. *Ibid.*, pp. 36-37.

186. For an in-depth analysis of Europe-wide radical rightist populists, see Cas Mudde, *Populist Radical Right Parties in Europe*, Cambridge, UK: Cambridge University Press, 2007, pp. 158-183.

187. For some clarity on the accession process for the Western Balkan states, see *ec.europa.eu/enlargement/enlargement_process/accession_process/how_does_a_country_join_the_eu/sap/index_en.htm.*

188. "Montenegro formally applies to join European Union," December 16, 2008 available from *web.archive.org/web/20090426011014/www.eubusiness.com/news-eu/1229351535.73.*

189. "EU Offers Serbia Candidacy For Membership," March 2, 2012, available from *www.rferl.org/content/european_union_serbia_membership/24502015.html.*

190. Available from *ec.europa.eu/enlargement/potential-candidate countries/bosnia_and_herzegovina/eu_bosnia_and_herzegovina_relations_en.htm.*

191. "Montenegro formally applies to join European Union," December 16, 2008, available from *web.archive.org/web/20090426011014/www.eubusiness.com/news-eu/1229351535.73.*

192. *Ibid.*

193. See Rosa Balfour and Corina Stratulat, "Enlargement to the Balkans: One Step Forward, Two Steps Back," *Commentary*, European Policy Center, December 12, 2011, available from *www.epc.eu/documents/uploads/pub_1379_enlargement_to_the_balkans.pdf.*

194. Eva Gross and Alessandro Rotta, "The EEAS and the Western Balkans," IAI Working Papers Vol. 11, No. 15, June 2011, Rome, Italy: Istituto Affari Internazionali, available from *www.iai.it/pdf/DocIAI/iaiwp1115.pdf.*

195. "Serbia Politics: Judgment Day Deferred," *The Economist Intelligence Unit*, October 13, 2001, available from *viewswire.eiu.com.*

196. Bojana Barlovac, "West Balkans May Face Renewed Schengen Visas," *Balkan Insight,* June 12, 2012, available from *www.balkaninsight.com/en/article/serbia-may-face-visa-free-suspension.*

197. Gerald Knaus and Alexandra Stiglmayer, "Balkan Asylum Seekers and the Spectre of European Hypocrisy," *EU Observer,* October 4, 2011, available from *euobserver.com/7/113807.* Between December 2009 and December 2010, the EU lifted Schengen visa requirements for Albania, Bosnia-Herzegovina, Macedonia, Serbia, and Montenegro on condition that the five states implement reforms in passport security and border controls, curtail illegal migration, and combat organized crime.

198. Saso Ordanoski, "The Story of Macedonian Populism: 'All We Want is Everything'," in Abramowitz *et al.,* p. 108.

199. An account of regional cooperation sponsored by the EU and its limited achievements can be found in Charalambos Tsardanidis, "EU and South-Eastern Europe: From Asymmetrical Inter-Regionalism to *Dependencia* Sun-Regionalism?" *Southeast European and Black Sea Studies,* Vol. 11, No. 4, December 2011, pp. 489-509.

200. Bedrudin Brljavac, "Bosnia's Failing Europeanization Progress—Who is to Blame?" *Transconflict,* October 11, 2011, available from *www.transconflict.com/2011/10/bosnias-failing-europeanization-process-who-is-to-blame-110.*

201. Based on conversations with government officials and analysts in Sarajevo and Banja Luka, Bosnia-Herzegovina in June 2011.

202. Florian Bieber, "Building Impossible States? State-Building Strategies and EU Membership in the Western Balkans," *Europe-Asia Studies,* Vol. 63, No. 10, December 2011, pp. 1783-1802.

203. Apaula M. Picketing, "The Constraints on European Institutions' Conditionality in the Western Balkans," *Europe-Asia Studies,* Vol. 63, No. 10, December 2011, pp. 1939-1944.

204. See Zorana Brozovic, "Territorial and Border Demarcation Disputes in the Western Balkans. Case Study: Territorial and Boundary Disputes Between Serbia and Kosovo," Belgrade, Serbia: Center for Security Policy, December 2011.

205. See "Eulex and Rule of Law in Kosovo," Prishtina, Kosova: Balkan Policy Institute, May 2011, available from *policyinstitute.eu/eng/*.

206. Bojana Barlovac, "Kosovo, Serbia Reach Deal on Regional Representation," *Balkan Insight*, February 24, 2012, available from *www.balkaninsight.com/en/article/kosovo-serbia-reach-deal-on-regional-representation*.

207. Ivan Krastev, "The Balkans and the EU After Mladic," *Survival*, Vol. 53, No. 4, London, UK, August 2011, p. 16, available from *dx.doi.org/10.1080/00396338.2011.603559*.

208. See Gulnur Aybet and Florian Bieber, "From Dayton to Brussels: The Impact of EU and NATO Conditionality on State Building in Bosnia and Herzegovina," *Europe-Asia Studies*, Vol. 63, No. 10, December 2011, pp. 1911-1937.

209. James Pettifer, "The Western Balkans: Returning Instability," *Balkans Series*, Vol. 11, No. 02, Defence Academy of the United Kingdom, July 1, 2011, available from *www.da.mod.uk/publications/library/balkan-series/20110701-Balkans-Series-11-01.pdf/view*, p. 4.

210. For an analysis of security in the light of NATO reductions, see Foreign Policy Club, "NATO Presence in Kosovo. Implications of KFOR Troop Reduction and Essential Political Processes," Prishtina, Kosova, May 2011, available from *www.fes-prishtina.org/wb/media/Publications/2011/Nato%20Presence%20in%20Kosovo%20(English).pdf*.

211. Available from *www.b92.net/eng/news/politics-article.php?yyyy=2012&mm=01&dd=04&nav_id=78118*.

212. Gerry J. Gilmore, "U.S. Peacekeeping Operations Continue in Balkans," July 13, 2004, available from *www.defense.gov/news/newsarticle.aspx?id=25716*.

213. EUFOR Fact Sheet, available from *euforbih.org/index.php?option=com_content&view=article&id=15&Itemid=134.*

214. Kori Schake, "U.S. Retrenchment is Right and Overdue," in Tomas Valasek, ed., *All Alone? What U.S. Retrenchment Means for Europe and NATO*, February 2012, London, UK, p. 24.

215. Azinovic, Bassuener, and Weber, pp. 74-75.

216. "Rogozin: You Can Join NATO Without Kosovo," *B92*, Belgrade, Serbia, February 6, 2010.

217. "Russian Diplomat Criticized For Speech At Serbian Opposition Rally," *Radio Free Europe/Radio Liberty*, November 2, 2011, available from *www.rferl.org/content/russian_diplomat_criticized_for_speech_at_serbian_opposition_rally/24379085.html.*

218. Stephen Blank, "A Scandal in Serbia," *Eurasia Daily Monitor*, Vol. 8, Issue 201, November 1, 2011, available from *www.jamestown.org/single/?no_cache=1&tx_ttnews%5Btt_news%5D=38606&tx_ttnews%5BbackPid%5D=7&cHash=5ed28d12e8cb1ebc3d67739c4688b842.*

219. "Serbia Politics: Serbia to be Included in Russia's Eurasia Union?" *The Economist Intelligence Unit*, March 1, 2011, available from *viewswire.eiu.com*, based on a report by Aleksander Palic, "East Calling Serbia," on the website of the Serbian newspaper *Vecernje Novosti* on February 19, 2011.

220. "Kosovo Serbs Want Russian Citizenship," *The Moscow Times*, November 27, 2011, available from *www.themoscowtimes.com/news/article/kosovo-serbs-want-russian-citizenship/448714.html#ixzz1ex1rUSEF.*

221. *Ibid.*

222. Available from *lafamillia.hubpages.com/hub/Nikolic_and_Putin.*

223. Radovan Vukadinovic, "The Geopolitics of Orthodox Solidarity," *Democracy and Security in Southeastern Europe*, Atlantic Initiative, Sarajevo, Bosnia, Vol. 1, Issue 4/5, December 2010, p. 21.

224. "Macedonia Relaxes Visa Rules for Russians," *The Moscow Times*, June 18, 2012, available from *www.themoscowtimes.com/news/article/macedonia-relaxes-visa-rules-for-russians/460449.html#ixzz1zODeq3aW*.

225. "Russia, Macedonia to Establish Free Trade Zone," *FOCUS News Agency*, June 21, 2012.

226. Biljana Lajmanovska, "Macedonia Eyes Kosovo, Albania as Gas Markets," July 11, 2012, *Southeast European Times*, available from *www.setimes.com/cocoon/setimesxhtml/en_GB/features/setimes/features/2012/07/11/feature-03*.

227. Vukadinovic, p. 23.

228. See Drasko Duranovic, "The Challenges and Dangers of Intimacy," *Democracy and Security in Southeastern Europe*, Atlantic Initiative, Sarajevo, Bosnia, Vol. 1, Issue 4/5, December 2010, pp. 31-34.

229. Government Statistical Office, Podgorica, Montenegro, July 12, 2011, available from *monstat.org/userfiles/file/popis2011/saopstenje/saopstenje(1).pdf*.

230. For details on Russian purchases throughout Eastern Europe, see Janusz Bugajski, *Expanding Eurasia: Russia's European Ambitions*, Washington, DC: CSIS Press, 2008, pp.104-132.

231. Ivana Jovanovic, "Russia's Biggest Bank Enters the Region," *Southeast European Times*, August 6, 2012, available from *setimes.com/cocoon/setimes/xhtml/en_GB/features/setimes/features/2012/08/06/feature-02*.

232. For economic statistics, consult IMF, *The World Economic Outlook*, September 2011, available from *www.imf.org/external/pubs/ft/weo/2011/02/weodata/index.aspx*; United Nations Conference on Trade and Development (UNCTAD), World Investment Report 2011, available from *www.unctad-docs.org/files/UNCTAD-WIR2011-Full-en.pdf*; and CIA, *The World Factbook*, available from *https://www.cia.gov/library/publications/the-world-factbook/index.html*.

233. Reports on democratization in the West Balkans can be found in Rosa Balfour and Corina Stratulat, "The Democratic Transformation of the Balkans," Brussels, Belgium: European Policy Center, *Issue Paper,* No. 66, November 2011, available from *www.epc.eu.*

234. Arch Puddington, *Freedom in the World 2012: Selected Data From Freedom House's Annual Survey of Political Rights and Civil Liberties,* Washington, DC: Freedom House, January 19, 2012.

235. For a comprehensive report on shortcomings in institution building that impede accession to the EU, see *The Western Balkans: Between the Economic Crisis and the European Perspective,* Sofia, Bulgaria: Institute for Regional and International Studies, September 2010, available from *www.irisbg.org/files/The%20Western%20Balkans.pdf.*

236. See Jovan Teokarevic, "Ten Years of Post-Milosevic Transition in Serbia: Problems and Prospects," in Abramowitz *et al.,* pp. 68-69, available from *www.iss.europa.eu/publications/detail/article/the-western-balkans-and-the-eu-the-hour-of-europe/.*

237. Simonida Kacarska *et al.,* "20 Years After 1991: The Tale of Two Generations," *Opinion Paper, Supporting Policy Development Paper Series,* 1/2012, Belgrade, Serbia: European Fund for the Balkans, available from *balkanfund.org/spd/?p=127.*

238. *World Report 2012,* New York: Human Rights Watch, available from *www.hrw.org/world-report-2012/bosnia-and-herzegovina;* and *Country Reports on Human Rights Practices for 2011,* Washington, DC: Department of State, Bureau of Democracy, Human Rights and Labor, available from *www.state.gov/j/drl/rls/hrrpt/humanrightsreport/index.htm?dynamic_load_id=186322.*